ENGINEERING, INFORMATION AND AGRICU
THE GLOBAL DIGITAL REVOLUTION

PROCEEDINGS OF THE 1ST INTERNATIONAL CONFERENCE ON CIVIL ENGINEERING, ELECTRICAL ENGINEERING, INFORMATION SYSTEMS, INFORMATION TECHNOLOGY, AND AGRICULTURAL TECHNOLOGY (SCIS 2019), 10 JULY 2019, SEMARANG, INDONESIA

Engineering, Information and Agricultural Technology in the Global Digital Revolution

Editors

Aria Hendrawan

Universitas Semarang, Indonesia

Rifi Wijayanti Dual Arifin

Research Synergy Foundation, Bandung, West Java, Indonesia

CRC Press
Taylor & Francis Group
Boca Raton London New York

CRC Press is an imprint of the
Taylor & Francis Group, an **informa** business

A BALKEMA BOOK

Published by:
CRC Press/Balkema
P.O. Box 447, 2300 AK Leiden, The Netherlands
e-mail: Pub.NL@taylorandfrancis.com
www.crcpress.com – www.taylorandfrancis.com

First issued in paperback 2021

ISBN 13: 978-1-03-224213-2 (pbk)
ISBN 13: 978-0-367-33832-9 (hbk)

DOI: https://doi.org/10.1201/9780429322235

Typeset by Integra Software Services Pvt. Ltd., Pondicherry, India

Visit the Taylor & Francis Web site at
http://www.taylorandfrancis.com

and the CRC Press Web site at
http://www.crcpress.com

Library of Congress Cataloging-in-Publication Data

Engineering, Information and Agricultural Technology in the
Global Digital Revolution – Hendrawan & Wijayanti Dual Arifin (eds)
© 2020 Taylor & Francis Group, London, ISBN 978-0-367-33832-9

Table of contents

Engineering, Information and Agricultural Technology in the
Global Digital Revolution – Hendrawan & Wijayanti Dual Arifin (eds)
© 2020 Taylor & Francis Group, London, ISBN 978-0-367-33832-9

Foreword

The Industrial Revolution 4.0 is a vital activity that has as of late been presented in Indonesia. The point is the change of modern assembling through digitalization and abuse of the capability of new advancements. In this manner, the Industrial 4.0 generation framework is adaptable and enables items to be balanced exclusively and explicitly. The idea of the Industrial Revolution 4.0 isn't restricted to assembling organizations yet additionally incorporates a total worth chain from supplier to client and all business capacities and administrations. The Industrial Revolution 4.0 is an Internet of Things specialization that is applied to the assembling/mechanical condition. This expects ongoing information gathering which prompts issues in taking care of and breaking down enormous information and digital security (Rojko A, 2017).

The improvement of this industry depends on information that keeps on developing, causing changes in the realm of training. This is a test for Universities, particularly the University of Semarang, to create techniques with the goal that institutional abilities, learning, asset capabilities must have the option to help this improvement. Along these lines, Semarang University through a worldwide meeting, in particular, the Semarang International Conference Series (SICS) 2019 Science Series: Civil Engineering, Electrical Engineering, Information Systems, Information Technology, and Agricultural Technology Conference had the topic "Reacting to Digital Revolution". In it contains research articles that examine the effect and arrangements in managing the advancement of the computerized transformation. We trust this gathering can answer the requirements of the network and the creating business today.

Aria Hendrawan
Rifi Wijayanti Dual Arifin

Engineering, Information and Agricultural Technology in the
Global Digital Revolution – Hendrawan & Wijayanti Dual Arifin (eds)
© 2020 Taylor & Francis Group, London, ISBN 978-0-367-33832-9

Organizing Committee

General Chair
Dyah Nirmala Arum Janie

General Co-Chair
Aria Hendrawan
Anna Dian Savitri
Subaidah Ratna Juita
Tatas Transinata
Hendrati Dwi Mulyaningsih

Conference Coordinator
Santi Rahmawati
Ani Rachmawati
Febrialdy Hendratawan

Conference Support
Almas Nabili Imanina

Information and Technology Support by Scholarvein Team

Engineering, Information and Agricultural Technology in the
Global Digital Revolution – Hendrawan & Wijayanti Dual Arifin (eds)
© 2020 Taylor & Francis Group, London, ISBN 978-0-367-33832-9

Scientific Review Committee

Editors:

Aria Hendrawan
Universitas Semarang, Indonesia

Taufiq Dwi Cahyono
Universitas Semarang, Indonesia

Andi Kurniawan Nugroho
Universitas Semarang, Indonesia

Supari
Universitas Semarang, Indonesia

Sri Budi Wahjuningsih
Universitas Semarang, Indonesia

Rohadi
Universitas Semarang, Indonesia

Scientific Reviewers:

Nurul Aini Osman
INTI International University, Malaysia

Essien Akpan Antia-Obong
Newcastle University, United Kingdom

Titin Winarti
Universitas Semarang, Indonesia

Kiran Singh
Indian Institute of Technology, Roorke, India

Chandrakant Sonawane
Symbiosis Institute of Technology, Deemed University, India

Haslina
Universitas Semarang, Indonesia

Quanjin Ma
Universiti Malaysia Pahang, Malaysia

Rohadi
Universitas Semarang, Indonesia

Faaizah Shahbodin
Universiti Teknikal Malaysia Melaka (UTeM), Malaysia

Nasir Jamal
Wuhan University of Technology, China

Mudjiastuti Handajani
Universitas Semarang, Indonesia

Shalini Shen
International Institute for Population Sciences (IIPS), India

Sri Budi Wahjuningsih
Universitas Semarang, Indonesia

Engineering, Information and Agricultural Technology in the
Global Digital Revolution – Hendrawan & Wijayanti Dual Arifin (eds)
© 2020 Taylor & Francis Group, London, ISBN 978-0-367-33832-9

Smart transportation that is integrated with information and communication technology will accelerate the formation of the smart city

M. Handajani
Universitas Semarang, Semarang, Indonesia

ABSTRACT: Transportation in developed countries usually uses smart transportation that is connected with information and communication technology in a government city. In Indonesia the transportation system still seldom uses smart transportation and public transportation is still mixed with private vehicles. Private vehicles account for 85% of total vehicles. So it is necessary to improve the city transportation systems by improving public transport services. The purpose of this article is to attract private passengers to mass transit, by increasing the proportion of the transportation system that is integrated with information and communication technology. Implementing smart transportation will affect the development of a city for the better, which essentially creates a smart city. The use of mass transportation will reduce fuel consumption and increase fuel efficiency, making city transportation low energy, low pollution, and environmentally friendly. The integration of smart transportation with information and communication technology supported by the intelligent transportation system will accelerate the formation of smart cities because information about transportation can be accessed and carried out more easily, effectively, and efficiently.

Keywords: Smart City, Smart Transportation, Intelligent Transportation System, Information and Communication Technology

1 INTRODUCTION

Almost every day people use transportation, so they want to do so in a practical way. Technology is rapidly developing, bringing changes to transportation that are easy, effective, and efficient, namely by implementing smart transportation. A smart city is a city that deals with problems, provides solutions, then makes improvements (Chandra, 2016). The purpose of this article is to attract private passengers to mass transit, by increasing the proportion of the transportation system that is integrated with information and communication technology. For this reason, it is necessary to develop an in-depth transportation system by installing passenger information systems to change public transportation from manual systems and switch them to digital technology. This makes people enter a modern lifestyle that cannot be separated from electronic devices, which are easier, more effective, and efficient. If technology is utilized properly, it will provide a lot of conveniences and positive impacts on the community, which brings economic and sociocultural consequences and changes travel behavior. Transportation that is environmentally sound and that has been connected with information and communication technology in city governance can improve the efficiency of public services. Implementing smart transportation integrated with information and communication technology will accelerate our society toward smart cities.

2 RESULT AND DISCUSSION

2.1 *Land used*

Current conditions for the development of urban land use in Indonesia are still developing horizontally (next to right – left) while in cities of developed countries land use is developing vertically. This could be addressed by a compact land-use system (compact city). A compact city will produce low/efficient fuel/capita consumption. With low fuel consumption, a city will use low energy and generate low pollution. The relationship of population density in major cities in the world with the use of fuel is presented in Figure 1. According to Handajani (2011), almost all cities in Java are at the bottom, meaning that the use of BBM/capita in developed countries is much higher than the use of BBM/capita in cities on Java.

2.2 *City governance*

Currently, people prefer to use private vehicles rather than public mass transportation; this is due to lack of time efficiency and the inconvenience of traveling to a distantly located bus stop. The implementation of the expected smart transportation, however, is environmentally sound, safe, affordable, effective and efficient. To achieve a smart city, the development of smart transportation must also be balanced by city governance that is integrated with information and communication technology. For example, cities can clear sidewalks of street vendors and install information boards that provide public information, traffic information, advertisements, breaking news, and other interesting information to provide comfort for pedestrians. The smart pedestrian can be seen in Figure 2.

Provision of smart shelter services can also increase interest in public mass transportation. With the installation of sensor ticketing, solar cells, city real-time information, and virtual stores in each shelter as shown in Figure 3, the public, especially prospective public transportation passengers, will feel safer and more comfortable.

Passenger information systems installed in each shelter will certainly increase public interest in mass transportation. Each shelter has a unit installed to monitor bus positions in real time,

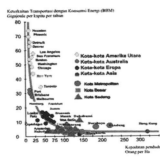

Figure 1.   Relationship of population density in cities in the world to fuel use.

Source: Handajani (2016)

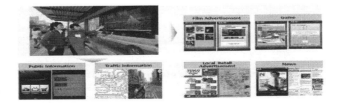

Figure 2.   Smart pedestrian.

Source: Sutisno (2016)

2

Figure 3.   Provision of smart shelter services.
Source: Sutisno (2016)

arrival and departure times, and the number of passengers per bus as applied in several developing countries. The passenger information system that has been implemented for the city of Semarang is presented in Figure 4.

With smart parking, the system can reduce congestion. Smart parking is integrated with information and communication technology so that prospective users can find available parking, which can be provided in vertical or horizontal forms such as depicted in Figures 5(a) and (b).

2.3   Smart transportation

Smart transportation is environmentally friendly, safe, convenient, inexpensive, and modern. Smart transportation is very influential in the creation of smart cities. According to Wei-Hasun, Tseng, and Shieh (2009), an intelligent transportation system is a system that can help users get information related to transportation and traffic, such as real-time traffic density. Included in the intelligent transportation system are: (1) an area traffic control system (ATCS), and (2) an integrated ticketing system.

2.4   Area traffic control system (ATCS)

An area traffic control system (ATCS) is an information and communication technology–based traffic control system that aims to optimize the performance of the road network. Such efforts include giving green lights to priority vehicles (fire engines, ambulances, VVIPs, etc.). The steps needed to support ATCS activities include:

Figure 4.   Information system in Semarang.
Source: Handajani (2018)

   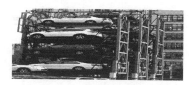

Figure 5.   (a) Horizontal smart parking system.   Figure 5.   (b) Vertical smart parking system.
Source: Google.com (2018)

1. Improvement of the traffic signal system. The term APILL is very foreign to most people because it is rare for people to use it in daily transportation activities. But the public knows it as a traffic light. Indeed, APILL is a piece of traffic equipment that is used as a signal in the form of red, yellow, and green lights placed at a crossroads to regulate traffic. This APILL system still needs to be upgraded and integrated with technology so that it can monitor the status of the controller junction, extend and shorten green lights when necessary, and correlate between controllers as in Figure 6.
2. Design of a rapid transit bus (BRT) priority system. The BRT system uses an APILL receiver at crossroads to capture the signal installed on the BRT, which can be connected wirelessly. The APILL will remain green when the BRT crosses the intersection. The following BRT configuration design priority that has been associated with APILL can be seen in Figure 7.
3. Installation of variable message signs (VMS). Variable message signs (VMS) are instructions sent through the operational control room or traffic management center in the form of an LED board, such as the one shown in Figure 8.
4. Implementation of video surveillance. Video surveillance in the form of 24-hour live video can monitor traffic conditions so that the general public can access them via the Internet in the form of digital video recording (DVR). Video surveillance via closed-circuit television (CCTV) can be seen in Figure 9.

China already employs a system that uses sensors to get the number of passengers on a bus; each passenger can be monitored on the bus and the control center will automatically record the information in a database (Zhang, 2013). The information system is very important for passengers because it is one of the efforts to withdraw passengers from private vehicles to public mass transportation, by providing arrival information in real time (Swati, 2013). The passenger information system installed at the bus stop can provide information such as time of arrival, time of

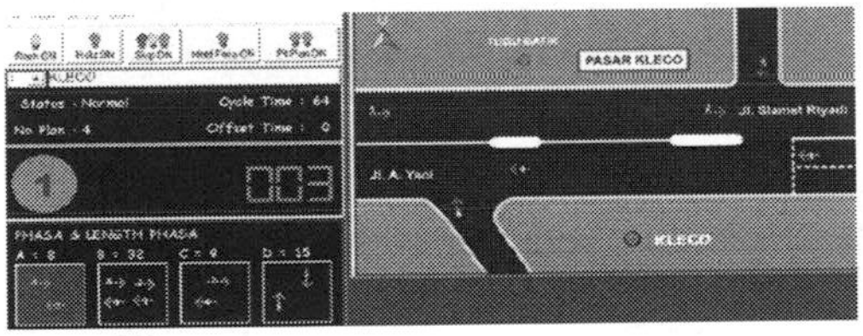

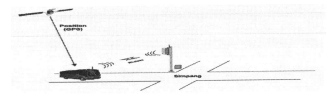

Figure 6.   APILL monitoring system.
Source: Hidayati (2017)

Figure 7.   Configuration of rapid transit bus (BRT) connected with APILL.
Source: Hidayati (2017)

Figure 8.   Variable message sign (VMS).
Source: Sutisno (2016)

Figure 9.   Design of video surveillance (CCTV).
Source: Fitria (2016)

departure, number of passengers, and position of the bus as applied in the cities of developed countries. In Indonesia, information systems began at the bus stop, precisely in the city of Semarang, to monitor bus trips and passengers. The monitoring system is placed in the Semarang City Office of Transportation as a central control center, and a passenger information system is installed in each shelter so that prospective passengers can monitor the state of the bus voluntarily and can know the time of departure. Monitoring and passenger information systems consists of three main components: (1) bus unit: a GPS device tracker, microcontroller module, and GSM module to find out data (speed, position, destination bus number, and station) and emergency signals; 2) client-side application: a unit that displays and monitors bus tracking in real time for prospective passengers; 3) central control center (CCU): the CCU functions to monitor the state of the bus in real time and includes receivers, emergency receivers, and microcontrollers that are simultaneously accessed by prospective passengers. The monitoring and passenger information system can be seen in Figure 10.

2.5   Integrated ticketing system

An integrated ticketing system is an email communication system that users can access if they encounter problems and questions; users can quickly contact the support team from the integrated ticketing system and receive relevant responses. The integrated ticketing system is focused on efficient travel time, and prospective passengers are provided with safe and comfortable facilities to make ticket payment transactions easily. In Jakarta, Indonesia, automatic ticket machines have also been applied to mass rapid transit (MRT) services. An image of the integrated ticketing system in Paris and of the automated ticket machines in Jakarta, which were implemented in 2019, can be seen in Figures 11(a) and 11(b).

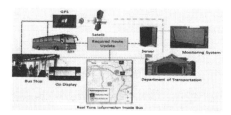

Figure 10.   Configuration of the monitoring and passenger information system.
Source: Handajani (2018)

Figure 11.   (a) Travel and ticket management in Figure 11. (b) Automatic ticket machine in
Paris.   Jakarta.
Source: Handajani (2011)

5

Transactions at the gate utilize integrated circuit (IC) cards and electronic payments. Where the IC card is one card and meets all needs, it is often called a smart card. Smart cards can provide convenience, efficiency, and security for users. The MRT station in Jakarta has also provided facilities that make the ticket payment system easier and encourage smart transportation in Indonesia. By using transit access pass (TAP) smart cards at the gate, prospective passengers purchase tickets according to their destination. Payment at the gate using a smart card is shown in Figure 12(a), and the service facilities of MRT stations in Jakarta are presented in Figure 12(b).

The shape and size of smart cards are the same as the shape and size of conventional credit cards, but smart cards work differently than credit cards. In a smart card usually, an embedded microprocessor serves to increase the security of user data. If users want to read and write data to the random access memory (RAM) embedded in the smart card, the microprocessor must be connected with a computer host and a card reader. The microprocessor embedded in a smart card is presented in Figure 13.

Use of smart cards can be expanded to include various transactions in transportation, health care, schools, bill payments, parking, tolls, and so on. By trading using a smart card that can also store personal data, transactions will be easier, safer, more convenient, and more efficient. The configuration of uses of smart cards can be seen in Figure 14.

The foregoing analysis shows that the establishment of a smart city is influenced by the development of smart transportation integrated with information and communication technology and supported by an intelligent transportation system and city governance.

Figure 12.   (a) Payment at the gate with a smart card.

Source: Handajani (2011)

Figure 12.   (b) Payment at the gate with a smart card in Jakarta.

Figure 13.   The microprocessor embedded in smart cards.

Source: Google.com (2018)

Figure 14.   Configuring usages of smart cards.

Source: Hastono (2016)

3 CONCLUSION

Organizing smart transportation will help societies accelerate toward smart cities. This will be beneficial for passengers and transport operators alike because smart transportation uses a system that is more modern, safe, comfortable, inexpensive, easy, affordable, and environmentally sound. Buses and other vehicles that are part of the smart transportation system should have their own road section, and smart transportation innovations can be supported by the private sector and the government.

ACKNOWLEDGMENTS

The author would like to express his gratitude to RISTEKDIKTI for providing the opportunity and support in making the HIBAH of Higher Education Applied Research (PTUPT), Transportation, University of Semarang.

REFERENCES

Chandra, H. 2016. Development strategy: Smart city and challenges for urban communities. *Journal of Strategy and Business, 4*(2).

Fitria, Z. 2016. Traffic density information system based on Raspberry PI PC Board. *National Journal of Electrical Engineering, 5*(1).

Handajani, M. 2011. The influence of urban transportation system in Java on city fuel. *Proceedings of the 4th ASEAN Civil Engineering Conference.*

Handajani, M. 2016. Fuel saving solution towards sustainable transportation. Inauguration of Professor of Civil Engineering, Faculty of Engineering, University of Semarang.

Handajani, M. 2018. Monitoring and passenger information system trans Semarang Bus Corridor VI.

Hastono, B. 2016. Implementation of intelligent transportation system. *FGD. ITS.*

Hidayati, Q. 2017. Control of traffic lights with vehicle detection using the method blob detection. *JNTETI, 6*(2).

Sutisno, T. 2016. The core role in realizing smart city based on an intelligent transportation system.

Swati, C. 2013. Implementation of real-time bus monitoring and information systems. *International Journal of SRP, 3*(5).

Wei-Hasun, L., Tseng, S. S., & Shieh, W. Y. 2009. Collaborative real-time traffic information generation and sharing framework for intelligent transportation systems. *Information Sciences, 180,* 62–70.

Zhang, Y. 2013. Intelligent urban public transportation system oriented to passenger travel and implementation method thereof. *Guangzhou (CN) U.S.*

Engineering, Information and Agricultural Technology in the
Global Digital Revolution – Hendrawan & Wijayanti Dual Arifin (eds)
© 2020 Taylor & Francis Group, London, ISBN 978-0-367-33832-9

Intergrading earth & disaster science to enable sustainable adaptation & mitigation

Magaly Koch
Center for Remote Sensing, Boston University, Boston, USA

ABSTRACT: Coastal cities worldwide are facing the enormous task of becoming resilient to physical, social and economic challenges; in addition to challenges due to climate change. However, our understanding of the major factors contributing to multiple coastal hazards is often incomplete due to the complexity of the problem, data scarcity and lack of research infrastructure for conducting transdisciplinary research. This paper advocates for the establishment of national and international networks of experts to better understand how coastal environments respond to such challenges by combining field experience with cutting edge geospatial technology and data analytics.

1 COASTAL GEOHAZARDS

Many large cities are immensely vulnerable to natural disasters because they are located in areas prone to geohazards. This is especially the case with coastal areas in Southeast Asia where the population is exposed to a variety of hazards, such as earthquakes, volcanic eruptions, floods, cyclones, droughts, and landslides (Arthurton, 1998; Djalante *et al.*, 2017). The combination of high population concentration, inadequate infrastructure and poor socioeconomic conditions results in high vulnerability to hazards impact. Natural disasters are likely to be made worse by global warming and climate change.

In developing countries, such as Indonesia, flooding is one of the frequent natural hazards. Indonesia is at the third place of the most vulnerable countries due to flood hazard in Asia, after China and India. In coastal urban areas, e.g. Semarang in Central Java, seawater tidal flooding is enhanced by land subsidence and is a major threat for city development (Abidin *et al.*, 2013; Rudiarto, Handayani and Sih Setyono, 2018). Developing countries tend to be hit particularly hard by such issues, and have a strong need for research and innovation to better manage and adapt to extreme events.

However, our understanding of the major factors contributing to geohazards is often incomplete due to the complexity of the problem, data scarcity and lack of research infrastructure for conducting transdisciplinary research. Meanwhile, the impact of rapid urban and economic development in hazard-prone coastal areas is one of the new emerging research topics. There is an urgent need for investigating how the natural system responds to anthropogenic forcing and vice versa, especially in tectonically dynamic, climatically sensitive and highly populated regions of the World such as Southeast Asia.

2 INTEGRATED APPROACH FOR COASTAL HAZARDS MITIGATION

There is a general consensus that the present changes of coastal zones results from several forcing factors (natural and anthropogenic). Thus assessing the impacts of present and future coastal hazards requires an understanding of the complex interactions between geoenvironmental, biophysical and socioeconomic systems. This can be best achieved by an integrated approach that includes research on both land and sea dynamics to identify natural and anthropogenic factors, their relative influences and related consequences. An integrated approach is the only way of pursuing the monitoring of such complex natural and human systems under various present and future climatic scenarios. Such approach should include among others:

- Measuring surface deformation on coastal cities
- Modeling accurately the coastal flooding from tides and storm surges
- Evaluating morphological changes due to sedimentary processes and/or human interventions
- Assessing tsunamis and geo-hazards risks (including landslides and land subsidence)
- Monitoring coastal marine ecosystems health and productivity (coral reefs, mangroves)
- Monitoring land cover/use changes and their impact on coastal communities
- Assessing coastal communities' access to fresh water and public health facilities
- Investigating linkage between flooding, waste management and public health

These studies are needed to mitigate or to adapt to changing conditions (e.g. coastal flooding, landslides, coral diseases, land subsidence), and prepare for potential crisis. However, an effort must be brought to effectively integrate the different types of studies (climatic, hydrodynamic, ecosystemic, anthropogenic etc.) in order to predict the integrative effects of global change in coastal zones.

Furthermore, research is required on how to transform scientific knowledge into a decision support system to enable a variety of stakeholders (farmers, fishers, local/regional government, international organizations etc.) to understand short and long term consequences of land use changes and urban development. Innovative tools and methods are need to better harness big data, accurately model and predict current and future scenarios, improve data visualization techniques and promote effective communication to various stakeholders (Goswami *et al.*, 2018). What is needed is a generalizable framework to transform environmental science into policy action through rapid computational scenario analysis and decision-making procedures on short to long-term time scales in order to enable problem solving solutions. In other words, a framework should be envisioned that effectively integrates earth and disaster science to engineer sustainable adaptation and mitigation measures.

Such research efforts require linkages between various disciplines and research facilities to enable data collection and validation, model development and calibration, and data analysis within an integral framework. This is best achieved by cooperation between networks of researchers, academic institutions, government agencies and other stakeholders in the domestic as well as international arena.

3 GLOBAL NETWORKS

Global networks of earth observing systems are exponentially growing, facilitating the collection and distribution of global data from space and ground based instruments to scientists around the world. At the same time access and sharing of information has increased as well as the sharing of computational resources. The technology of data sharing and processing has advanced significantly in the last decade. What is needed is an infrastructure that enables forging new relationships between national and international research networks to foster collaborative actions to tackle complex system issues.

Supersites or natural laboratories are examples of such collaborative initiatives that provide access to earth observation and geophysical data for key priority sites affected by earthquakes, volcanic eruptions and other forms of geohazards. Other similar initiatives include the "100 Resilient Cities (100RC)" initiative funded by the Rockefeller Foundation (www.100resilientcities.org). This program aims to help cities worldwide to be resilient against physical, social and economic challenges, in addition to challenges due to climate change. Semarang (Indonesia) is one of the cities selected to participate in the initiative. This city exemplifies the multiple threats affecting society, economy, environment, and infrastructure. Solving or at least mitigating the multifaceted problem is a challenge, requiring research in many disciplines and involving multiple experts and stakeholders.

Thus there is a clear need for fostering networking of experts and users across countries and regions to promote and facilitate access and sharing of data, tools, and research infrastructure for incubating new technologies and problem solving ideas. Leveraging resources and knowledge across disciplinary and regional boundaries requires funding mechanisms to enable such activities. Funding programs for establishing such networks would need to be broad and inclusive, encompassing a range of disciplines and participating countries.

For example big data from space is providing massive spatiotemporal earth observation data to detect, monitor, and assess in almost real time geohazards threating coastal areas especially those with high population density and infrastructure. However, the complexity of the problem demands coordinated research activities by multiple experts from different disciplines and with access to research facility sharing. This could be promoted by funding programs that enable 1) research exchange visits; 2) data collection and sharing facilities; 3) joint field work; 4) laboratory access; 5) training workshops; and 6) international seminars/workshops to incubate new ideas and disseminate results.

4 FINAL REMARKS

New emerging data-intensive fields require multidisciplinary approaches to tackle complex issues as well as transformative improvements in graduate education. Enabling students and post-docs to perform their training and research in multiple laboratories and/or field sites enriches their research experience and prepares them better for future career paths. Global issues affect more than a single country or region, and cannot be resolved without international collaboration.

In summary, an integrated approach will lead to a better understanding of complex systems as a whole, discovery of problem solving solutions, improved assets utilization, and transfer of knowledge across disciplines and regions.

REFERENCES

Abidin, H. Z. *et al.* (2013) 'Land subsidence in coastal city of Semarang (Indonesia): characteristics, impacts and causes', *Geomatics, Natural Hazards and Risk*. Taylor & Francis, 4(3), pp. 226–240.

Arthurton, R. S. (1998) 'Marine-related physical natural hazards affecting coastal megacities of the Asia–Pacific region–awareness and mitigation', *Ocean & Coastal Management*. Elsevier, 40(1), pp. 65–85.

Djalante, R. *et al.* (2017) 'Introduction: Disaster Risk Reduction in Indonesia: Progress, Challenges, and Issues', in *Disaster Risk Reduction in Indonesia*. Springer, pp. 1–17.

Goswami, S. *et al.* (2018) 'A review on application of data mining techniques to combat natural disasters', *Ain Shams Engineering Journal*. Elsevier, 9(3), pp. 365–378.

Rudiarto, I., Handayani, W. and Sih Setyono, J. (2018) 'A regional perspective on urbanization and climate-related disasters in the northern coastal region of Central Java, Indonesia', *Land*. Multidisciplinary Digital Publishing Institute, 7(1), p. 34.

Engineering, Information and Agricultural Technology in the
Global Digital Revolution – Hendrawan & Wijayanti Dual Arifin (eds)
© 2020 Taylor & Francis Group, London, ISBN 978-0-367-33832-9

Using satellite images and GIS for proposing practical mitigation measures against floods affecting Semarang City, Indonesia

A. Gaber

Geology Department, Faculty of Science, Port Said University, Egypt

ABSTRACT: Urban flooding due to high-intensity rainfall along coastal Semarang City is a major problem every winter. Such urban flooding cannot be avoided, but it can be controlled. Thus, it is necessary to extract real information regarding the flood events as well as to understand the landscape of Semarang using an integrated approach of remote sensing and a geographic information system (GIS) together with field measurements to propose reliable mitigation measures. In this work, the digital elevation model (DEM) of Semarang was used to automatically extract all the morphometric and hydrological parameters. In addition, different satellite images (optical and radar) were used to map the land surface features and to estimate the subsidence rate that affects the city and changes the land slope. Moreover, geological, structural, and groundwater maps were integrated in a GIS database. Consequently, several measures were proposed to mitigate the flooding. These measures are: construct prevention dams at the up streams, striate the meandering rivers, dredge the water channels to increase discharge, increase the river levee height, inject the excess water into a productive groundwater aquifer after performing quantitative hydrological modeling, construct a separate drain network (totally different than the sewer system), and, finally, install artificial water infiltration and attenuation systems in the flat areas. The locations and design of all these mitigation measures were mapped in a GIS format.

1   INTRODUCTION

Semarang City is located in a low-relief area at the footwall of highly elevated mountains located to the south, and it is considered a downstream coastal city (Figure 1). Semarang receives an average precipitation rate of 2,500 mm/year (Figure 2). Therefore, Semarang has witnessed an intensive urban expansion during the past few decades with increasing numbers of built-up areas (Figure 3). The topography, the weather, and urban activities along the coastal lowland areas mean that Semarang suffers from urban flooding, especially in the winter season from December to March, with peaks during December, January, and February. Figure 4 shows the expansion trend of the built-up areas in Semarang from 1972 to 2009.

This work aims at using different space-borne satellite images with the aid of a geographic information system (GIS) to extract information about the psychographics of Semarang as well as information from real flood events to propose practical flood mitigation measures. In addition, the land subsidence that affects the city is estimated by utilizing a differential interferometric synthetic-aperture radar (InSAR) methodology that uses radar sentinel-1 satellite images.

Figure 1.   Location map shows Semarang City at the footwall of a mountain.

Figure 2.   Average annual rainfall in Semarang.

Figure 3.   Built-up areas of Semarang.

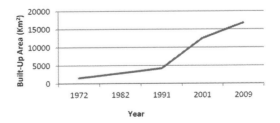

Figure 4.   The expansion trend of the built-up areas in Semarang from 1972 to 2009.

2 MATERIALS AND METHODS

In this work, several sources of data covering Semarang were collected and integrated in a GIS database. A total of 12 Sentinel-1 radar images covering the period from February to December 2018 was downloaded and processed using a satellite-based augmentation system (SBAS) interferometry technique to estimate land subsidence along Semarang (Bürgmann, Rosen, & Fielding, 2000; Gabriel, Goldstein, & Zebker, 1989). Figure 5 shows the connection graph of the used 12 Sentinel-1 images and the selected master image of February 2018.

3 RESULTS AND DISCUSSIONS

Figures 6 to 10 show the land subsidence rate from February until December. The northeastern part of Semarang shows the highest subsidence rate, which might be related to the huge urban expansion as well as to an intensive groundwater withdrawal along these areas. In addition, this location has the thickest alluvium deposits along the city's coastal areas.

Moreover, based on the interpretations of the satellite images, two different mitigation measures were proposed: upstream and downstream mitigation measures. The upstream mitigation measures include building dams and straightening the meandering streams. The downstream measures comprise intensive and continued dredging of the drains, increasing the height of the levee, and constructing drains on both sides of the road networks. We recommend investigating the pattern of a real flood event using the different satellite images in order to understand very well the effects and the trends of such flood events and to propose practical and visible mitigation measures.

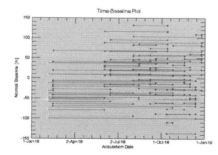

Figure 5. The connection graph of the used 12 Sentinel-1 images.

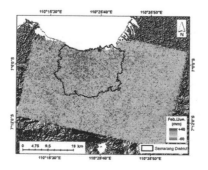

Figure 6. The estimated land subsidence from February to June 2019.

13

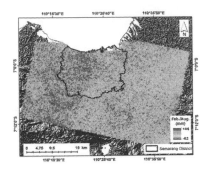

Figure 7. The estimated land subsidence from February to August 2019.

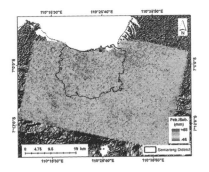

Figure 8. The estimated land subsidence from February to September 2019.

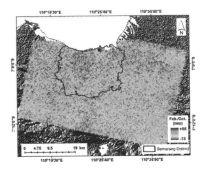

Figure 9. The estimated land subsidence from February to October 2019.

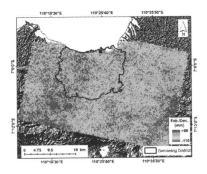

Figure 10. The estimated land subsidence from February to December 2019.

14

REFERENCES

Bürgmann, R., Rosen, P. A., & Fielding, E. J. 2000. Synthetic aperture radar interferometry to measure Earth's surface topography and its deformation. *Annual Review of Earth and Planetary Sciences, 28*(1), 169–209.

Gabriel, A. K., Goldstein, R. M., & Zebker, H. A. 1989. Mapping small elevation changes over large areas: Differential radar interferometry. *Journal of Geophysical Research: Solid Earth, 94*(B7), 9183–9191. Available from the Wiley Online Library at https://agupubs.onlinelibrary.wiley.com/doi/abs/10.1029/JB094iB07p09183.

Engineering, Information and Agricultural Technology in the
Global Digital Revolution – Hendrawan & Wijayanti Dual Arifin (eds)
© 2020 Taylor & Francis Group, London, ISBN 978-0-367-33832-9

The computerization of Faarfield and Comfaa for the PCN value analysis of the runway 13-31 Ahmad Yani International Airport Semarang in 2018

W. Hermanto, B. Priyatno, P.A.P. Suwandi & A. Sutami
Universitas PGRI Semarang, Semarang, Indonesia

ABSTRACT: The Airport is an important infrastructure in transportation activities in the world, especially in Indonesia, which is an archipelagic country. In Indonesia air transportation is a supporting factor for economic, social, cultural, industrial and tourism developments. This has led to the importance of the development of aircraft the fleets that are professional and have international standards. To analyze the thickness of the runway according to FAA standards, the ACN-PCN method is the only method determined by ICAO which refers to the FAA AC No 150/5320-6E provisions regarding the Airport Pavement Design and Evaluation and KP. No. 93 of 2015. With the completion of the 13-31 Runway International Airport Ahmad Yani Semarang overlay at the end of December 2018, it is necessary to recalculate the Runway Ability to support Aircraft Load Capacity using FAARFIELD and COMFAA so that the value of the 13-31 PCN Runway can be identified can be published.

Keywords: PCN Value, Airport, runway, Overlay

1   INTRODUCTION

Ahmad Yani International Airport Semarang is an airport managed by PT. Angkasa Pura I (Persero). Semarang Ahmad Yani International Airport continues the improvement to support the increase of the flight operations from year to year as seen from the addition of frequencies and flight lines and the increase in the number of passengers and goods (cargo)

With the growing flight operations at Ahmad Yani International Airport, Semarang, it should be noted that the runway's ability needs to remain excellent and always ready to use. One of the ways to maintain the runway's ability is to carry out the overlay work for runway 13-31 to increase the value of the previous PCN (Pavement Classification Number), which is PCN 54 F/D/X/T.

2   PCN ANALYSIS

Based on the growing number of the aircraft type operating in service of passengers and goods, the increasing of the runway's ability to service landings and flights require special attention in terms of planning, design, implementation, and maintenance. This is done to ensure the implementation of safety and security of flight operations according to the declaration of the Republic of Indonesia Law No. 1 of 2009 concerning Aviation and Air Transportation Director General of Regulation No. KP.93 of 2015 concerning Technical Standards and Operations of Civil Aviation Safety Regulations-section 139 (Manual of Standard CASR-Part 139) Volume I Airports (Aerodromes) (Kementerian Perhubungan, 2015).

Overlay work activities at Runway 13-31 Ahmad Yani International Airport, Semarang, have been completed. With the addition of thickness along runway 13-31, it is expected to increase the value of 13-31 PCN Runway. In general, the PCN value will be influenced by 2 (two) main factors, namely the pavement factor itself and the type of aircraft to be served by the pavement (FAA, 2014). From the pavement factor, the PCN value is measured from the subgrade value and the total thickness of the runway pavement structure which includes sub-base, base, and surface. Other factors that also influence are aircraft traffic load, aircraft type, aircraft weight at departure also the configuration and the pressure of the main landing gear. From the various types of aircraft that operate, it will be known the type of critical aircraft that has the highest CDF value (can be more than one) which has the most destructive potential for the pavement structure.

The increase of the PCN value of the runway 13-31 was also influenced by the difference in the overlay thickness on each runway segment due to the longitudinal and transverse geometric improvements whose minimum thickness will be analyzed using FAARFIELD Software. With this geometric improvement, the overlay thickness will be different for each slope which results in the difference in the equivalent thickness of each slope even in the same segment generated from the COMFAA spreadsheet. This equivalent thickness is used to be the input of COMFAA, along with data traffic and subgrade CBR values to increase the value of PCN also at once reducing the value of CDF during the age of pavement plan against critical aircraft ACN (Aircraft Classification Number).

After the overlay to the Runway 13-31, the following results are:

a. The dimensions of the runway 13-31 are 2560 meters long (after displacement of 120 meters) and 45 meters wide which is including in the 4D runway classification according to ICAO provisions. The runway 13-31 was built through the development stages from 1993 to 2012. This includes the runway construction, renewal and re-lining (leveling and overlaying) with flexible (asphalt) pavement types. Based on the analysis of the development stages of the runway 13-31, it is divided into 4 main segments where each of these segments has a material composition that differs in characteristics and thickness but stands in the lowest structure with a subgrade CBR value of 3%. the four segments are as follows.

Table 1. Distribution of runway segments 13-31.

Segment	Length	Year
1	805	Until 1993
2	925	1994
3	400	2005
4	430	2007

Source: researcher analysis

By using FAARFIELD software as stipulated in KP No. 93 of 2015 and FAA No AC 150/5320 6E, each segment is calculated based on the type and thickness of the pavement layer, and the type of aircraft operating (Perhubungan, 2015; FAA, 2018).

b. The computerized results show that the most critical types of aircraft are B737-800 and B737-900ER with ACN 51 values (Airplanes, 2005). Both are the types of aircraft that have the highest ACN value and most operate on runway 13-31. the Analysis of the pavement conditions after overlaying found that the runway 13-31 CDF values in the existing conditions are in the range of 0.00 to 0.0975, which indicates that with the existing aircraft operations, the runway is being able to support the operation of critical aircraft for service periods for the next 20 years.

Table 2. The Results of FAARFIELD's software analysis.

Segment	Existing thickness (mm)	CDF Value	Overlay thickness (mm)	Equivalent thickness (mm)
1	1285,0	0,000	142,7	1817,0
2	1285,0	0,000	145,6	1822,0
3	1355,0	0,000	177,0	1847,0
4	1135,0	0,0975	148,2	1690,0

Source: researcher analysis

c. From the average thickness of each segment and the CBR value that has been previously known, the initial analysis to determine the range of PCN runway 13-31 is done using a COMFAA spreadsheet to obtain the equivalent thickness of each slope of each runway segment 13-31. The results of this calculation also recommend the runway 13-31 pavement type.
d. For the security of flight operations, the PCN value of runway 13-31 can be set at 73 F/D/X/T, where this value exceeds the PCN value in the previously existing condition of 54 F/D/X/T.

Table 3. The PCN Value of Runway 13-31.

Segment	Existing thickness (mm)	Average Overlay thickness (mm)	PCN value per segment
1	1285,0	142,7	76 F/D/X/T
2	1285,0	145,6	75 F/D/X/T
3	1355,0	177,0	74 F/D/X/T
4	1135,0	148,2	73 F/D/X/T

3 SUGGESTION

After the PCN analysis of the runway 13-31 Ahmad Yani International Airport, the PCN value declaration can be planned and known even before the overlay activity is carried out. However, this is a count that can only be technically used as a reference. To maintain the safety and security of flight operations, the authors suggest that:

a. Carry out routine maintenance and supervision of runway 13-31 to ensure the implementation of security and safety of flight operations at Ahmad Yani International Airport Semarang.
b. The B737-900 ER is the aircraft with the highest ACN value served by runway 13-31. COMFAA Software takes into account the service period of 20 years (coverages) as many as 29,311 departures or approximately 4 departures per day, for this reason, it is necessary to re-analyze the CDF value and remaining life of pavement if there is an increase in the coming years during the pavement service.
c. The analysis in this paper is done theoretically. To prove the actual PCN value after overlaying it should be carried out a Heavy Weight Deflectometer (HWD) test to obtain PCN data analytically directly from the runway.
d. If the number of flights especially for critical aircraft changes in a significant increase, it is necessary to re-analyze the strength of runway 13-31 by using FAARFIELD and COMFAA software to determine the runway power capacity of 13-31 to the operation of the type of aircraft served.
e. The load that will be received by the runway pavement structure will be smaller than the load on the apron and taxiway (Sartono, 1992). This is because when parking at the apron, the aircraft can refuel which adds weight to reach MTOW conditions. While in taxiways, the aircraft will run slowly and reduce the weight of the aircraft itself before reaching the runway, hence it is necessary to do a separate calculation to calculate the PCN value of the ACN for the critical aircraft types both in the apron and taxiway.

f. With the original soil condition which is a former pond area, it is necessary to observe whether the runway pavement structure 13-31 has decreased or not for the anticipatory steps of airport management as part of evaluation and monitoring within the scope of Pavement Management System activities and makes it an advanced analysis which can complement this writing.

REFERENCES

Airplanes, B. C. (2005) '767 Airplane characteristics for airport planning (D6-58328)', *Seattle, WA: Author*.

FAA (2014) *Airport Pavement Design and Evaluation, Advisory Circular AC: 150/5320-6E,*. Washington. D.C.: U.S. Departement of Transportation.

FAA (2018) *Aircraft Characterictic, Apendix 1, Advisory Circular AC: 150/5300-13A*. Washington. D.C.: U.S. Departement of Transportation.

Kementerian Perhubungan (2015) 'KP. 93 tahun 2015 tentang Pedoman Teknis Operasional Peraturan Keselamatan Penerbangan Sipil bagian 139-24 (Advisory Circular Part 139-24) Pedoman Perhitungan PCN (Pavement Classification Number) Perkerasan Prasarana Bandar Udara', *Kementerian Perhubungan*.

Sartono, W. (1992) 'Airport Engineering', *Biro Penerbit, Keluarga Mahasiswa Teknik Sipil, Universitas Gadjah Mada, Yogyakarta*.

Engineering, Information and Agricultural Technology in the Global Digital Revolution – Hendrawan & Wijayanti Dual Arifin (eds)
© 2020 Taylor & Francis Group, London, ISBN 978-0-367-33832-9

The influence of compressive mortar geopolymer strength on the addition of carbit waste ash with a curing oven system

Anik Kustirini
Universitas Semarang, Semarang, Indonesia

Mochammad Qomaruddin
Nahdlatul Ulama Islamic University, Jepara, Indonesia

Diah Setyati Budiningrum
Universitas Semarang, Semarang, Indonesia

Imania Eka Andammaliek
Nahdlatul Ulama Islamic University, Jepara, Indonesia

ABSTRACT: With increasing infrastructure development, scientists are developing alternative materials for Portland cement; one of these is geopolymer. In this research, fly ash and carbide weld ash are used as binders or substitutes for cement by making cuboidal geopolymer mortar with a size of 5 cm x 5 cm x 5 cm with 150 specimens. The composition of the material used in this study is 50% fine aggregate ratio: 50% (activator + binder) with a binder ratio 60%:activator 40% (NaOH and Na_2SiO_3 compared 1:2). The test method carried out is compressive strength. The results of the research are: geopolymer mortar with pure fly ash has the highest compressive strength of 16.60 MPa, while geopolymer with 20% carbide weld ash has the highest mortar compressive strength of 17.35 MPa.

Keywords: geopolymer, compressive strength, fly ash, carbide ash, oven curing

1 INTRODUCTION

Concrete is a very popular material used in buildings, and it consists of coarse aggregates, fine aggregates, water, and Portland cement, which are the most important ingredients because they are all aggregates. However, when the cement production process occurs, the release of carbon dioxide (CO_2) gas into the air is proportional to the amount of cement produced (Davidovits, 1994), which can damage our environment.

a. Fly ash
   Fly ash is the remnants of coal combustion that flow from the combustion chamber. Fly ash, a very light, grayish powder, is an inorganic oxide material containing silica (SiO_2) of 58.20%.
   Fly ash is included in B3 waste with code D223 denoting the main pollution of heavy metals.
b. Carbide waste
   Carbide waste is the remainder of the carbide reaction to water, which can produce acetylene gas. Classified in the types of lime as stated by Zainal Abidin (1984), carbide waste has limestone properties for building materials in accordance with SII 0024-80 with the existence of two parameters, namely lower levels of CaO + MgO and high enough CO_2.
c. Sodium Hydroxide (NaOH)
   Sodium hydroxide serves to react the elements of Al and Si contained in fly ash so that it can produce strong polymer bonds. As an activator, sodium hydroxide must be dissolved

Table 1.　The content of fly ash.

Component	Percentage (%)
SiO_2	58.20
Al_2O_3	18.40
Fe_2O_3	9.30
CaO	3.30
MgO	3.90
K_2O	0.60
SO_2 Na_2O	0.07

Table 2.　Waste content of carbide.

Chemical composition	Content (%)
SiO_2	4.3
Fe_2O^3	0.9
Al_2O_3	0.4
CaO	56.5
MgO	1.7
SO_3	0.06
LOI	36.1

Figure 1.　Image of fly ash.

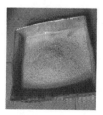

Figure 2.　Image of carbide ash.

first with water according to the desired molarity. This solution must be made and left at least 24 hours before use (Hardjito et al., 2004).

d. Sodium Silicate (Na_2SiO_3)

Sodium silicate can be made with two processes, namely a dry process and wet process. In the dry process, sand (SiO_2) is mixed with sodium carbonate (Na_2CO_3) or with potassium carbonate (K_2CO_3) at a temperature of 1,100–12,000°C. These reactions produce glass (cutlets) dissolved in water; with high pressure, the glass becomes clear, thick liquid. Sand (SiO_2) mixed with sodium hydroxide (NaOH) through the filtration process will produce pure silicates (Samlistiya, 2017).

e. Process Treatment/Curing

Curing is done after the casting in order to preserve the concrete during the hydration process. Concrete maintenance is strongly influenced by the temperature and humidity of the concrete itself; therefore, concrete treatments affect not only the strength of the concrete but also its durability. The use of effective treatment methods depends on the type of material used, the type of construction, and the expected concrete utilization. Two concrete treatment methods are used based on the temperature, namely normal maintenance and treatment at elevated temperature (Neville & Brooks, 1987).

2 METHODOLOGY

The authors examined geopolymer mortar with basic ingredients, namely fly ash and carbide weld ash. The authors used these basic ingredients because they aimed to utilize industrial waste that is of no economic value for concrete manufactures and to reduce the environmental impact of the accumulation of industrial waste itself. The mortar was made with materials from fly ash obtained from the waste of PLTU Tanjung Jati B and with carbide weld ash from the remaining metal welding from a carbide workshop located in Kalongan Jepara. The test specimens were made with several different comparisons of the percentage of fly ash and carbide ash and using comparisons (1FA: 0 carbide, 0.9 FA: 0.1 carbide, 0.8 FA: 0.2 carbide, 0.7FA: 0.3 carbide, 0.6 FA: 0.4 carbide) and alkali activator (8 M NaOH +Na_2SiO_3). After the test object was made and printed, it was removed from the mold and then treated or cured. Two treatment methods were employed: the first was to leave the test object at room temperature (31–32°C), and the second was to treat the test object by steam curing in an oven with a temperature difference of 100°C and 200°C for a full hour. The test object was left for a specified time, then its compressive strength was tested with a compressive strength machine.

3 RESULTS AND DISCUSSION

3.1 *Results of material characteristics test*

Analysis was carried out to determine the physical and chemical characteristics of the materials used in this study. Analysis of physical properties consisted of testing the cleanliness of the material in regards to organic matter, sludge/precipitation, and wash mud; material filter analysis was conducted as well. The chemical and physical properties were examined using an SEM-EDX analysis for carbide welding waste. The material analysis carried out in this study included analysis of sand material, fly ash, and carbide weld ash.

Concerning the levels of organic matter in the sand, the color of the experimental results should indicate that the sand is not older than the color of the comparison substance, namely NaOH. If the color of the results indicates an age older than that of the NaOH, it will adversely affect the mortar itself. From experiments carried out, the color of the test content of the organic substance was brownish yellow, which can be compared with the color of the tintometer.

Table 3. Analysis of the cleanliness of sand against organic matter.

Experiments	Number of	Units of
high sand + mud	158	Cc
high sand	121	Cc
height of mud	23	Cc
Color caused	Brown yellow	

Table 4. Analysis of the sand content of sludge.

Experimental	Number of	Units of
sand and mud	146	mm
High sand (H)	140	mm
Mud height (h)	6	mm
Mud content =	4	%

Sand slurry levels were good: the sand sludge levels were lower than 5%. If the content of sand sludge is higher than 5%, then the sand must be washed before use. From the research experiment, it was found that the sand sludge was 4%, so the conditions of the sand used in the study met the requirements.

3.2 Filter analysis

Before conducting the filter analysis, the materials and tools collected included sand, sieves to filter the sand (size 4.75, 2.36, 1.18, 0.6, 0.3, 0.15), an oven to dry the sand (if in SSD), and digital scales.

The steps are:

a. Dry the sand first, if it in the SSD state, using the oven.
b. While waiting for the sand to dry, prepare a filter by arranging filters in order of size from the largest filter above to the smallest below/pan at the bottom.
c. Remove the sand from the oven, cool it, and weigh out 1,000 gr. of sand.
d. After weighing, put the sand into the filter lid and place the filter arrangement on the shaker. The filter device is shaken for 15 minutes.

3.3 Initial time binding

The initial tie time is the time needed to harden from the reaction of a binder to an activator and to become a geopolymer paste. The binder must be stiff enough to hold the load. There is a setting time at the time of hardening or binding between fiber binders.

1. Binding time of the initial mix of design 0% carbide
   The initial binding time is 172.8 minutes or approximately 3 hours. The final binding time, which is a decrease, indicates that the number 0 mm occurs in 240 minutes.
2. Binding time of the initial 10% mix
   The addition of 10% carbide ash can affect the initial binding time. The initial tie time with the addition of 10% carbide reaches 81 minutes and has begun to harden.
3. Binding time of the initial mix design 20% carbide
   The addition of 20% carbide greatly affects the binding time, which is 72 minutes.
4. Binding time of the initial mix design 30% carbide
   The addition of 30% carbide ash reduces the binding time to 20 minutes. The initial binding time is 25 minutes. When making a mix of designs with the addition of 30% carbide before 25 minutes, stir all the material that has begun to stick, so that the binder and activator mixture attached to it can immediately be poured into the mold so that it does not dry before printing.
5. Binding of the initial mix design 40% carbide
   The addition of 40% carbide results in a very short binding time of 16 minutes. With the addition of 40% carbide ash to sample making, the mortar should not be mixed for more than 20 minutes, because the binder mixture can harden before it is poured into the mold.

3.4 *Compressive strength*

This section explains the effects on compressive strength of age, mix design, and temperature. Before discussing compressive strength, the composition of geopolymer mortar is again outlined, which consists of sand material, carbide weld ash, and fly ash. Alkaline solutions are NaOH (8 MOL) and Na_2SiO_3.

In this study, carbide weld ash was made into several mix designs with variations in the addition of carbide weld ash by 0%, 10%, 20%, 30%, and 40% to the weight of fly ash because it aims to determine the highest compressive strength against the effect of additional carbide ash. However, everything must be studied beforehand so that it meets the requirements. The results of the compressive strength test at the ages of 7, 14, and 28 days from the trial 5 cm x 5 cm x 5 cm cube are presented in the form of tables and graphs later in this article.

The following are the results of testing mortar compressive strength:

Mortar compressive strength for age at 100% fly ash (Figure 3)

At 7 days and room temperature (31–34°C), the mortar has compressive strength worth 8.27 MPa, which increases at 100°C by 0.0.741% (14.40 MPa), but decreases at 200°C by 0.125% (12.8 MPa). At 14 days and room temperature (31–34°C), the mortar has a compressive strength of 14.67 MPa, which decreases at 100°C by 0.023% (16.60 MPa), but increases at 200°C by 0.023% (14.67) MPa. At 28 days and room temperature (31–34°C), the mortar has a compressive strength of 15.38 MPa, which increases at 100°C by 0.079% (16.60 MPa), but decreases at 200°C by 0.070% (15.50 MPa).

Mortar compressive strength for age at 90% fly ash (Figure 4)

At 7 days and room temperature (31–34°C), the mortar has a compressive strength of 11.60 MPa, which increases at 100°C by 0.252% (14.53 MPa), but decreases at 200°C by 0.062% (13.67 MPa). At 14 days and room temperature (31–34°C), the mortar has a compressive strength of 11.87 MPa, which increases at 100°C by 0.282% (15.30 MPa), but decreases at a 200°C by 0.02% (15.00 MPa). At 28 days and room temperature (31–34°C), the mortar has a compressive strength of 13.47 MPa, which increases at 100°C by 0.274% (17.17 MPa), but decreases at 200°C by 0.099% (15.62 MPa).

Mortar compressive strength for age at 70% fly ash (Figure 5)

At 7 days and room temperature (31–34°C), the mortar has a compressive strength of 12.00 MPa, which increases at 100°C by 0.116% (13.40 MPa) and increases at 200°C by 0.044% (14.00 MPa). At 14 days and room temperature (31–34°C), the mortar has a compressive strength of 14.00 MPa, which increases at 100°C by 0.076% (15.07 MPa) and increases again at 200°C by 0.041% (15.70 MPa). At 28 days and room temperature (31–34°C), the mortar

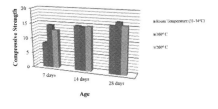

Figure 3.   Graph of temperature variation at 100% fly ash for age.

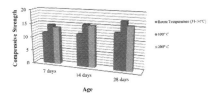

Figure 4.   Graph of temperature variation at 90% fly ash for age.

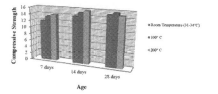

Figure 5.   Graph of temperature variation at 70% fly ash for age.

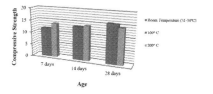

Figure 6.   Graph of temperature variation at 60% fly ash for age.

has a compressive strength of 15.00 MPa, which increases at 100°C by 0.043% (15.65 MPa), but decreases at 200°C by 0.052% (14.87 MPa).

Mortar compressive strength for age at 60% fly ash (Figure 6)

At 7 days and room temperature (31–34°C), the mortar has a compressive strength of 11.80 MPa, which does not experience changes at 100°C but increases at 200°C by 0.177 % (13.90 MPa). At 14 days and 32°C, the mortar has a compressive strength of 13.47 MPa, which does not change at 100°C and increases at 200°C by 0.039% (14 MPa). At 28 days and room temperature (31–34°C), the mortar has a compressive strength of 15.75 MPa, which decreases at 100°C by 0.009% (15.6 MPa) and decreases again at 200°C by 0.089% (14.32 MPa).

4   CONCLUSION

From the results of this research, the following can be concluded:

a. In addition to pure fly ash, which has the highest compressive strength of 16.60 MPa, the addition of 20% carbide weld ash has the highest mortar compressive strength of 17.35 MPa.
b. The addition of carbide weld ash into the waste treatment system using (curing) different mixes achieves the result that in each treatment (curing) temperature, the optimum compressive strength occurs only at 100°C. However, at 200°C, carbide ash will reduce the compressive strength of the mortar.
c. The more variations in the addition of carbide ash, the greater the effect of the speed of binding time on the binder. The lower the percentage of carbide ash, the longer the initial binding time.

REFERENCES

ASTM Committee C 191-04. 2003. Standard test method for time of setting of hydraulic cement by vicat needle.
ASTM Committee C 39-04a and AASHTO T22-151. 2007. Standard test method for compressive strength of cylindrical concrete specimens.
Budiarto, A., & Budi, C. 2007. Effect of karbit waste and fly ash on mortar strength. Surabaya: Final Project of S1 Thesis in Civil Engineering at Petra University.

Davidovits, J. 1994. Properties of geopolymer cements. In *First International Conference on Alkaline Cements and Concretes* (pp. 131–149). Kiev: Scientific Research Institute on Binders and Materials.

Duxson, P., Provis, J. L., Lukey, G. C., Mallicoat, S. W., Kriven, W. M., & Van Deventer, J. S. J. 2005. Understanding the relationship between the composition of geopolymers, microstructure and mechanical properties. *Colloids and Surfaces A: Physical and Chemical Aspects and Techniques, 269* (1–3), 47–58. doi: 10.1016/j.colsurfa.2005.06.060

Hardjito, D. et al. 2004. Factors influencing the compressive strength of fly ash-based geopolymer concrete. *Civil Engineering Dimension, 6*(2), 88–93.

Neville, A. M., & Brooks, J. J. 1987. *Concrete Technology*. London: Longman Scientific & Technical.

Samlistiya, N. 2017. Study of utilization of carbide waste and fly ash in geopolymer paste. Surabaya: Final Project Diploma Advanced Civil Engineering Program Level of Department of Building.

Shofi'ul Amin, M., Diky, F., Januarti Eka, P., & Triwulan. 2013. Alwa aggregate potential as a basic material for geopolymer concrete made from Sidoarjo mud. Civil Engineering ITS Surabaya. Surabaya.

Subekti, S. 2009. Strong resistance to press paste geopolymers 8 moles and 12 moles of molarity against Agersifta NaCl. Surabaya: National Technology Application Seminar Regional Infrastructure.

Engineering, Information and Agricultural Technology in the
Global Digital Revolution – Hendrawan & Wijayanti Dual Arifin (eds)
© 2020 Taylor & Francis Group, London, ISBN 978-0-367-33832-9

Implementing the augmented reality map to classify the damaged roads

Siti Asmiatun, Nur Wakhidah & Astrid Novitaputri
Universitas Semarang, Semarang, Indonesia

ABSTRACT: Land transportation is one transportation type widely used by many people in Indonesia. Land transportation is considered having the highest accident risks. Traffic accidents frequently become the problems due to the drivers' negligence and lack of disciplines to obey the traffic signs, as well as the perforated road conditions. Road detection is one most recently discussed topic in the research studies since public institutions, such as hospitals, police stations, schools, and etc. are connected by the roads. Many previous studies have discussed the implementation of augmented reality. However, the implementation of algorithm in decision making has not yet been developed. Therefore, researches on road detection have become one important topic to gradually develop. This study has implemented the Naive Bayes method to classify the damaged roads using the augmented reality map. The research data was taken from the accelerometer application in the previous researches [1]. This research has eventually resulted in the augmented reality map application to classify both minor and major damaged roads.

Keywords: Road, Detection, Augmented Reality, Map

1 INTRODUCTION

Information related to roads is one strategic and important issue in urban areas. Road detection is one most recently discussed topic in research studies because public institutions, such as hospitals, police stations, schools, and etc. are connected by the roads (Acar and Bayir, 2015). Therefore, a research on road detection become on important topic to gradually develop.

In our previous study, we discussed the detection of road conditions using the Z-Diff algorithm. The implementation method to filter the data was obtained from the smartphone's accelerometer sensor. The experimental results from the eastern Semarang area have showed that there were 281 perforated road points with the accuracy of 79% [4]. To develop a further research, we use a method combining the pothole detection and z-diff algorithm with the resulted accuracy of 74% in identifying the road surface conditions (Asmiatun, Wakhidah and Novita Putri, 2018). The detection of road conditions are further developed using the Naive Bayes method with the accelerometer sensor data with the resulted accuracy of 95.5% on rapid miner (Putri, Asmiatun and Wakhidah, 2018). So, it can be concluded that the Naive Bayes method is one best method to implement in this study.

This study has implemented the Naive Bayes method to classify the damaged roads using the augmented reality map. The research data was taken from the accelerometer application in the previous researches. A research on the augmented reality map has been conducted by Seungjun Kim resulting in the concept of car navigation display system. The roads were displayed on the car windscreen inside part using the augmented reality technology with virtual test-bed navigation parameters and distance virtually combined with GPS to facilitate the information related to the traffic accidents (Kim and Dey, 2009). Similarly, a research conducted by Nikolaos on patent publications has resulted in the concept of augmented reality navigation using 3D mapping objects and real-time video for the users' navigation instructions (Kokkas and Schubert,

2011). A research conducted by Zeljko on driving simulator by comparing the private road display (PSV) with the personal navigation devices (PND) augmented reality using a high-speed driving simulator. The PND Augmented Reality integrates the virtual navigation to the real-world screens by displaying it directly on the windscreen with head-up-display (HUD) that the drivers do no need to frequently see their cellphone screens while driving. Meanwhile, the PND Street View (PND SV) uses a sequence of road images and the surrounding roads as well as a virtual navigation route, to facilitate the users in making adjustments with the visual signs (Bifet and Frank, 2010). Referring to the previous researches, the technology of Augmented Reality Map may be developed for the transportation services, such as classifying the detection of damaged roads using the Naive Bayes algorithm.

2 METHODOLOGY

2.1 *Location*

The research objects were spread in several districts (in Indonesia known as *Kecamatan*) of East Semarang. The Damaged road data was obtained by conducting observations, with a data collection technique conducted by directly identifying the damaged roads in the research locations. The obtained data included the depth of the perforated roads, major shaky road areas, and speed.

2.2 *Naive Bayes*

Naive Bayes is a probabilistic classification algorithm type which uses the Bayes Theorem to classify the data. Naive Bayes may also be used for sentiment analysis, spam email detection, document categorization and many other fields (Bifet and Frank, 2010). The probabilities required by the Bayes Theorem may be calculated as follows:

$P (A \mid B) = P (B \mid A) \times P (A) \div P (B)$
Where A and B are events and P (B) is not equal to 0
P (A) is probability A while P (B) is probability B
P (A | B) is the probability of event A when event B is correct, while
P (B | A) is the probability of event B when event A is correct.

Table 1. Data training.

Depth of Perforated Road	ShakyRoad Area	Speed	Category
Small	Small	Fast	Minor Damage
Large	Small	Slow	Major Damage
Large	Small	Slow	Major Damage
Small	Small	Slow	Minor Damage
Small	Small	Slow	Minor Damage
Small	Small	Slow	Minor Damage
Small	Large	Slow	Major Damage
Small	Large	Fast	Major Damage
Small	Large	Fast	Major Damage
Small	Small	Fast	Minor Damage
Small	Small	Slow	Minor Damage
Large	Small	Fast	Minor Damage
Small	Small	Slow	Minor Damage
Small	Small	Slow	Minor Damage
Large	Small	Slow	Major Damage
Small	Small	Slow	Minor Damage
Small	Small	Slow	Minor Damage
Small	Small	Fast	Minor Damage
Small	Small	Slow	Major Damage
Large	Small	Slow	
Small	Small	Slow	?

Determining Criteria
Depth of Perforated Road = Small, Large
Shaky Road Area = Small, Large
Speed = Fast, Slow
Category = Minor Damage and Major Damage
Case
The above calculation results showed that the obtained record 20 classifications categorized into Minor Damage with the value of 0.59 when compared to the major damage with the value of 0.08.

3 RESULTS AND DISCUSSION

3.1 *Implementation of Augmented Reality Map with Naive Bayes on the Damaged Roads*

The flow of Augmented Reality Map with Naive Bayes on the Damaged Roads Damage started with the field study identification of the problems related to the damaged roads, then making the application using Z-Diff and Pothole Patrol method, data collection using the accelerometer application in East Semarang, classification of the collected data related to the damaged roads either major or Minor Damage using algorithm naive Bayes, and implementation of augmented reality map using the marker-less Tracking Map method.

3.2 *Results displaying the implementation of augmented reality map using unity 3D*

The results of implementation of augmented reality map using the Unity 3D Tools are displayed below. There is a menu of Broken Path Detection Pages and Classifications Using Naive Bayes. The results related to the damaged road Chart Application are shown in Figure 2.

Figure 3 shows the data training table of the damaged roads with the time field of x, y, and z, as well as Speed, Latitude, Longitude, and Road Category. In this view, there is a naive Bayes trial to calculate the category results, in which the data will be saved in the application database.

The results of classification calculation will display the results of Road chart with the category of major and minor damaged Roads classified using the Naive Bayes method as shown in the figure below.

Based on Figure 5, the classification results of the damaged roads are due to the calculation results of the naive Bayes method in which one damaged road category either major or minor damage will be displayed if one of the results is ≥ 0 but if the result is $== 0$ then the category will not be displayed. The image below is the final results of each category calculated using the Naive Bayes method as shown in Figure 5.

Figure 2.   Detection results of the damaged roads on the augmented reality map.

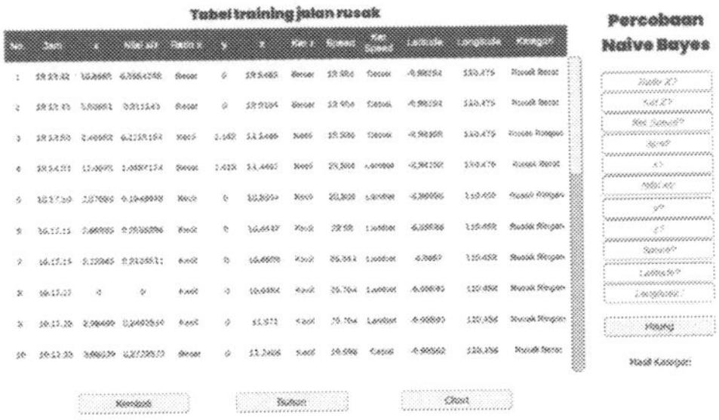

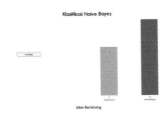

Figure 3. Training table of the damaged roads in augmented reality.

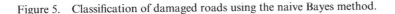

Figure 4. Chart of the damaged road category.

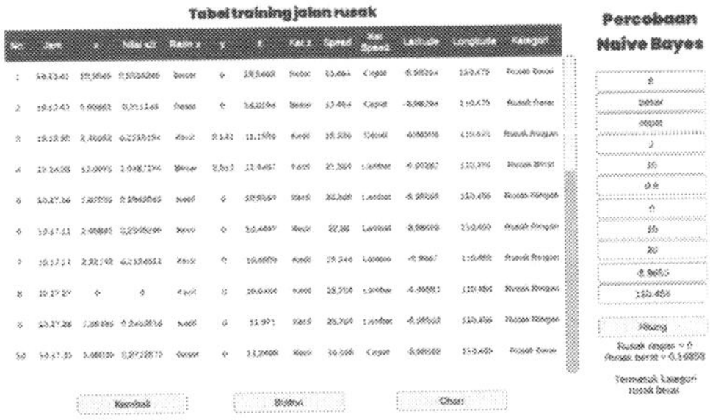

Figure 5. Classification of damaged roads using the naive Bayes method.

4 CONCLUSION

The conclusion of this research is the implementation of the augmented reality map application may be used to identify the map tracking location using the marker-less method serving as an information medium for the damaged roads. Naive Bayes classification method may be best used to classify both minor and major damaged roads. This application has weaknesses in determining and displaying the accuracy of both map longitude and latitude. If the longitude and latitude points slightly change, the displayed map accuracy results will be less optimal. For further researches, it is recommended to develop the accuracy of method results and compare them with the other classification methods.

REFERENCES

Acar, S. A. and Bayir, S. (2015) 'Road Detection Using Classification Algorithms.', *JCP*, 10(3), pp. 147–154.
Asmiatun, S., Wakhidah, N. and Novita Putri, A. (2018) 'Road Surface Automatic Identification System With Combination Pothole Detection Method and Z-Diff Method On Android Smartphone', in

Proceedings of the Joint Workshop KO2PI and the 1st International Conference on Advance & Scientific Innovation. ICST (Institute for Computer Sciences, Social-Informatics and . . ., pp. 39–46.

Bifet, A. and Frank, E. (2010) 'Sentiment knowledge discovery in twitter streaming data', in *International conference on discovery science.* Springer, pp. 1–15.

Kim, S. and Dey, A. K. (2009) 'Simulated augmented reality windshield display as a cognitive mapping aid for elder driver navigation', in *Proceedings of the SIGCHI Conference on Human Factors in Computing Systems.* ACM, pp. 133–142.

Kokkas, N. and Schubert, J. (2011) 'Method for the display of navigation instructions using an augmented-reality concept'. Google Patents.

Putri, A. N., Asmiatun, S. and Wakhidah, N. (2018) 'Klasifikasi Kondisi Permukaan Jalan Menggunakan Algoritma Naive Bayes', *Prosiding Semnas PPM 2018*, 1(1), pp. 556–561.

Briker server VoIP capability measurement over virtual local area network

W. Adhiwibowo & A. M. Hirzan
Universitas Semarang, Semarang, Indonesia

ABSTRACT: Telephone communication technology is still used everywhere. In a company, having a telephone is a must to facilitate communication with outside parties. Telephone communication technology has developed into a device with better connectivity and voice clarity known as Voice over Internet Protocol (VoIP). In using VoIP, a special device is needed for VoIP services using a computer network. A Briker server can bridge between VoIP data and a traditional phone line. In exchanging information, Briker server and network topology have limits on how much data traffic can flow. In this study, Briker's server was tested for its ability to provide VoIP services in a virtual local area network (VLAN). Mathematically, VoIP data traffic in this network can be calculated, so that it can be used to conclude whether the Briker server can handle this VoIP traffic.

1 INTRODUCTION

Basic communication devices such as telephones have become necessities in business. But if one employee accidentally nudges a telephone until it is lifted, it can swell a company's telephone bill.

Traditional telephone communication technology still uses a simple circuit that functions to connect speaker and microphone. This technology was later developed into a telephone device that uses Voice over Internet Protocol (VoIP) technology. This technology allows IP phone devices to communicate through a computer network so that the device will have an IP address. The VoIP technology can be utilized through various types of servers such as Trixbox, Elastix, Asterisknow, and Briker.

The Briker server is one of many VoIP Linux servers that runs on x86-64 architecture. Hence, in terms of performance, this server should provide better availability of VoIP communications. The Briker server also has a do-it-yourself feature through which users can customize the system to their needs. Most previous research that uses the Asterisk server run on embedded devices; the Briker server is available only on x86-64, also known as a personal computer (PC) platform. A PC has better processing power than embedded devices when using a VoIP service.

Most small offices and home-based businesses use a virtual local area network (VLAN) to reduce network costs, so it is quite important to know the load of a single trunk if VoIP is being implemented into the network. That's why proper research is needed regarding this matter.

This research tested communication performance using Briker servers on a VLAN network. An indicator that can be used to measure the performance of the Briker server in handling VoIP data traffic comprises a large flow of data and smooth communication using VoIP. It can be ascertained that this line will become heavy when various devices are actively communicating. From this, we can test the Briker server's ability to overcome VoIP traffic in this network configuration.

2 LITERATURE REVIEW

Previous research using different servers has also been done to find out how well VoIP is performing. The Asterisk server employing the Raspberry Pi 3 device can get a lot of good phone calls (Baliga, Chudasama, & Ambawade, 2017). Although the server can handle many telephone calls, 5% of calls are blocked due to the use of codec. Bandwidth and congestion also greatly affect the number of packets dropped by the system.

Scholars have also conducted research on the capacity of VoIP providers. Capacity testing carried out on the Asterisk server produced the following data (Costa, Lunes, Bordim, & Nakano, 2015):

1. With 40 active VoIP connections, the Asterisk server required a central processing unit (CPU) capacity of 15% to 20%. As many as 0% of the total 40 calls were blocked by the server.
2. The most numerous tests were 240 active calls, where the Asterisk server required 55% to 60% CPU capacity. Of the 240 calls, the Asterisk server blocked 29%.

Another study tested the performance and efficiency of the Asterisk server with Raspberry Pi devices and several other single-board computer devices that focus on the codecs used by VoIP (Menezes, Nogueira, Ordonez, & Ribeiro, 2018; Najihi, Mustika, Widyawan, & Najwaini, 2015). This research concluded that the Asterisk server can run properly without consuming too much memory, even though it is handling many active calls.

Research on CPU stress used VoIP data traffic on the Asterisk server (Ahmed & Mansor, 2009) with point-to-point network configuration, and produced data as follows:

1. Testing using advanced micro device (AMD) processors with a frequency of 1.5 GHz and without codec conversion produced 70 calls with a load of 60% on the processors.
2. An Intel processor with a frequency of 2.8 GHz produced 180 calls with a load of 60%.

Other studies experimenting with loading Asterisk servers (Pal, Triyason, & Vanijja, 2018) also used VoIP data traffic to test the robustness of the server.

Further studies also measured VoIP performance through wired and wireless mediums (Mohammed, 2016). These studies found a difference between jitter, end-to-end delay, throughput, and mean opinion score depending on the codec used.

3 METHODS AND APPROACHES

3.1 *Basic VoIP*

This research used VLAN networks as the basic configuration of VoIP networks, where three switches divide one network into three separate zones. The switch is linearly mounted like a tree trunk, with a router at the base of the network. Each of these switches is connected to several devices such as an IP phone, a server (only available on one switch), and one computer.

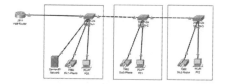

Figure 1. VLAN network logical topology.

3.2 *Mathematical model*

In order to test the performance of the Briker server, several things must be prepared:

1. Definition of the total call duration equation:

$$T(s) = (n \times (t^-)) \times 60 T(s) = (n \times (t^-)) \times 60 \tag{1}$$

where $T(s)$ is the total call duration, n is the total number of calls made, and t is the average call made in minutes.

1. Total call bandwidth:

$$T(bandwidth) = ((T(s) \times Bandwidth\ Rate)/8) \tag{2}$$

where $T(bandwidth)$ is the total bandwidth used during a call, and $T(s)$ is the total duration obtained from the previous formula then multiplied by the bandwidth rate of the calling codec.

2. Number of call packages:

$$N(packets) = (T(bandwidth))/65,507 \tag{3}$$

where $N(packets)$ is the total packet sent in VLAN, $T(bandwidth)$ is the total bandwidth used, and 65,507 is a defined maximum packet size (Postel, 1980).

3. Percentage of packet loss (Shi et al., 2008):

$$\left(\left(\sum \_(i=1)^{\wedge} n\ x\_i + x\_(i+1) + \ldots + x\_n \right) / N \right) \times 100\ \% \tag{5}$$

where x (one packet lost and zero packets received) becomes a packet sent from one client to another client, i is the beginning of the calculation using i, and n is the number of packages sent to the client.

4. Bandwidth memory:

$$Data(Byte/s) = (Clock \times Bus\ Wide \times N(channel) \times N(Interface)) \div 8 \tag{6}$$

where $Data(Byte/s)$ is the total data that random access memory (RAM) can store before going to the processor (Göbel, Elhossini, Chi, Alvarez-Mesa, & Juurlink, 2017), *Clock* is the clock rate of DRAM, *Bus Wide* is the width of the business data, $N(channel)$ reflects to, and $N(interface)$ is the number of interfaces within the computer system. This equation theoretically ignores the CPU bottleneck.

3.3 *Empirical calculation scenario*

Calculations were performed to find basic performance information from VoIP networks through VLAN that can be concluded hypothetically as to whether the Briker server can handle the VoIP data traffic through the VLAN network configuration.

This section examines four basic scenarios, each of them having a different number of calls. Two scenarios have one and three calls of different durations, and both scenarios are in ideal and loss situations. The rest of the scenarios have 50 and 100 calls of different durations, and both scenarios are in ideal and loss situations.

In order to serve these scenarios, the Briker server was set up with a 1 GHz 64-bit processor and DDR2 RAM 1 GB capable of processing 8 GB of data per second in parallel, under ideal conditions and ignoring overhead.

The results obtained from the empirical calculations were as follows:

Table 1. Empirical calculation results.

N-Call	Total Bandwidth (byte/s)		Threshold Bandwidth	
	G.723.1	G.771	G.723.1	G.771
1	4.68 MB/s	19.62 MB/s	✓	✓
3	23.4 MB/s	98.1 MB/s	✓	✓
50	421.2 MB/s	1.76 GB/s	✓	✓
100	1.35 GB/s	5.68 GB/s	✓	✓

According to this result, it can be concluded theoretically that the Briker server can handle 1, 3, 50, and 100 calls with a G.723.1 codec. The Briker server can also handle 1, 3, and 500 calls with a G.771 codec. However, the Briker server might have problems handling 100 active calls with a G.771 codec due to memory bandwidth limitations.

Hypothetically speaking, the Briker server can handle many active calls if provided with proper specifications. If the Briker server is equipped with proper specifications, then it can easily handle many active calls with a better VoIP codec.

When the server can manage VoIP well, then the next probable problem is the configuration of the VLAN network.

4 CONCLUSION

According to this research, it can be concluded that the VoIP capability of the Briker server over the VLAN network is as follows:

1. A higher number and longer duration of active calls will increase bandwidth usage and CPU server capabilities in providing VoIP services.
2. The selection of VoIP codecs affects VoIP voice quality, network bandwidth usage, and Briker server capabilities.
3. If the network system configuration and the server are in an ideal state, it will be easier to achieve more calls.
4. The Briker server processor with a clock speed of 1 GHz can handle fewer than 100 calls. If more than 100 calls occur, the server will be overloaded.

This research has benefits for those who wish to study the performance of single-trunk VLAN and VoIP using the Briker server theoretically or practically. This research can provide good advice for developers regarding the specification of the server and its topologies.

REFERENCES

Ahmed, M., & Mansor, A. M. 2009. CPU dimensioning on performance of Asterisk VoIP Pbx. *Proceedings of the 11th Communications and Networking Simulation Symposium*, 139–146.
Baliga, S., Chudasama, K., & Ambawade, D. 2017. Real-time performance evaluation and stability testing of Raspbx for Vowifi. *International Conference on Automatic Control and Dynamic Optimization Techniques, Icacdot 2016*, 1089–1095.

Costa, L. R., Nunes, L. S. N., Bordim, J. L., & Nakano, K. 2015. Asterisk Pbx capacity evaluation. *Proceedings of the 2015 IEEE 29th International Parallel and Distributed Processing Symposium Workshops, IPDPSW*, 519–524.

Göbel, M., Elhossini, A., Chi, C. C., Alvarez-Mesa, M., & Juurlink, B. 2017. A quantitative analysis of the memory architecture of FPGA-SOCS. *Lecture Notes in Computer Science (Including Subseries Lecture Notes in Artificial Intelligence and Lecture Notes in Bioinformatics)*, 10216 Lncs, 241–252.

Menezes, A. C., Nogueira, T. A., Ordonez, E. D. M., & Ribeiro, A. D. R. L. 2018. An approach to the performance and efficiency power analysis on embedded devices using Asterisk. *Journal of Computer Science, 14*, 1038–1052.

Mohammed, M. H. 2016. Performance analysis of VoIP over wired and wireless networks network implementation in Aden University. *International Journal of Research in Engineering and Technology, 5*, 325–330.

Najihi, A., Mustika, I. W., & Najwaini, E. 2015. Analisis kinerja IP PBX server pada single board circuit Raspberry PI. *Indonesian Journal of Computing and Cybernetics Systems, 9*(1), 89–100.

Pal, D., Triyason, T., & Vanijja, V. 2018. Asterisk server performance under stress test. *17th International Conference on Communication Technology (ICCT)*, 1967–1971.

Postel, J. 1980. RFC 768: User datagram protocol

Shi, L., Fapojuwo, A., Viberg, N., Hoople, W., & Chan, N. 2008. Methods for calculating bandwidth, delay, and packet loss metrics in multi-hop IEEE802.11 ad hoc networks. *IEEE Vehicular Technology Conference*, Spring 2008, 103–107.

Internet of Things model for public lighting

B.V. Christioko, A.F. Daru, W. Adhiwibowo & R. Prativi
Universitas Semarang, Semarang, Indonesia

ABSTRACT: The condition of public street lighting (*Penerangan Jalan Utama*) (PJU) at the moment does not meet the standard criteria set by the Perusahaan Listrik Negara (PLN), a situation caused by the behavior of people who install the street lighting incorrectly. Such activities may cause greater power consumption, which increases the monthly costs as well. As an effort to deal with the problem, it is necessary to create electronically controlled public street lighting (e-PJU). This e-PJU system is a combination of modern programming technology with microcontroller technology and Internet Protocol (IP)-based networking, which will be implemented on e-PJU devices. The model is also equipped with a monitoring and security system that can provide information on real-time lighting conditions, temperature, and the status of the lights. The e-PJU model will be integrated with Android for monitoring in real time, and e-PJU activity will be logged to facilitate information before conducting regular maintenance.

1 INTRODUCTION

Increasing road user activity, as well as developments in the construction and repair of public roads, have led to a demand for better public street lighting services. However, the current condition of public street lighting (*Penerangan Jalan Utama*) (PJU) in some regions still does not meet the standard criteria set by the Perusahaan Listrik Negara (PLN), due to the behavior of people who install street lighting illegally. Such activity causes uncontrolled, unmonitored power consumption.

In general, public street lighting systems use electricity funded with local government financing. Not only do regional and local governments finance the electricity bills but they also incur costs for the maintenance and replacement of identified devices that have exceeded the usage limit. Maintenance carried out by regional governments consists of two types, namely scheduled maintenance and unscheduled maintenance. Scheduled maintenance is the maintenance of devices that have passed the usage period while unscheduled maintenance is carried out when the device is damaged before the tool's usage period ends. If damage to the PJU cannot be detected from the start, it will disrupt daily community activities. In addition to the damage to the PJU, funding will increase due to theft of PJU devices.

Financing for PJU that is borne by regional governments is increasing from year to year along with regional growth and development. Therefore it is necessary to monitor the PJU so that the costs for PJU addition and maintenance of PJU can be controlled. As an effort to deal with this problem, it is necessary to create an electronic public street lighting system (e-PJU); this e-PJU system is a combination of modern programming technology with microcontroller technology and technology networking with Internet Protocol (IP), which will be implemented on PJU devices. The e-PJU model is also equipped with a monitoring and security system, where the system can provide information on real-time lighting conditions such as the location of the e-PJU, the temperature of the device, and the status of the light. In terms of security, this system is equipped with an account to access the device.

The e-PJU model will be integrated with Android-based applications as a medium for monitoring data in real time. In the application, there are also additional features such as an e-PJU activity data log, to make it easier for technicians to carry out regular maintenance.

2 LITERATURE REVIEW

2.1 *Previous research*

Research related to smart systems has been carried out by previous researchers. Research on the development of Internet of Things (IoT) technology has been applied to the health care system via a hybrid method that optimizes virtual machines. The hybrid method is a combination of a genetic algorithm (GA), particle swarm optimizer (PSO), and parallel particle swarm optimization (PPSO) used to build virtual machines (Abdelaziz et al., 2018). Researchers have also designed an IoT-based light-emitting diode (LED) light architecture using the Message Queuing Telemetry Transport (MQTT) method (Hu et al., 2018). Smart systems using indoor lighting models have been estimated to save up to 40% of electrical energy (Mahajan & Markande, 2017). Research with Zigbee Gateway technology has also been conducted using LEDs for highway lighting. The LED technology can save 68% to 82% of electrical energy (Shahzad, Yang, Ahmad, & Lee, 2016).

2.2 *Internet of Things*

Internet of Things (IoT) is a concept that uses the Internet as the main infrastructure network (Shahid & Aneja, 2017). In this case, IoT can also be interpreted as using the Internet to connect devices, where devices include information such as metadata (Bari, Mani, & Berkovich, 2013). Applications from IoT can be classified into a variety of uses, as seen in Figure 1 (Asghar, Negi, & Mohammadzadeh, 2015).

2.3 *Message Queuing Telemetry Transport (MQTT)*

Message Queuing Telemetry Transport (MQTT) is a simple message delivery protocol designed for limited bandwidth devices employing machine-to-machine (M2M) telemetry and used for communication of IoT devices. Initially, MQTT was designed to connect the telemetry system; it is quickly becoming one of the main protocols for deploying IoT.

This MQTT uses the publisher-subscriber principle so that one of the devices has the task of making an announcement about the configured topic while another device receives the announcement.

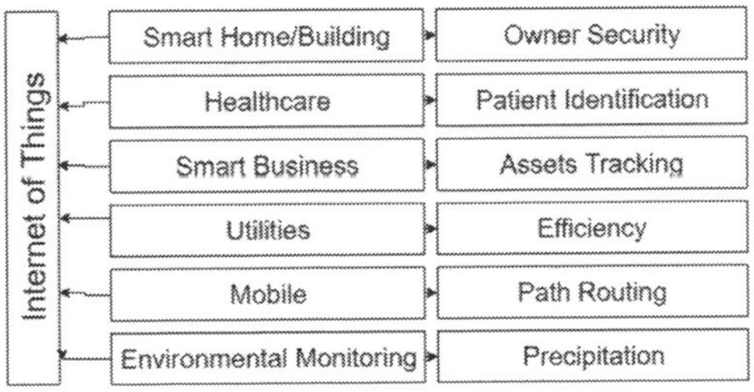

Figure 1. Internet of Things functions.

3 RESULTS AND DISCUSSION

3.1 *Research method*

This research used a special method that can only be employed in IoT device development. Internet of Things design methodology comprises nine main system development steps (Figure 2).

3.2 *Result*

The results of the research can be described as follows:

3.2.1 *Information model of e-PJU*
The model is seen in Figure 3, which explains that the main devices of this system are streetlights and power cables. Both of these devices are connected to the e-PJU system located on the highway, and each has a "status" attribute. Information obtained from these two devices indicates whether they are on or off. The e-PJU system monitors by using its sensors to get information from around the highway.

3.2.2 *Device* and *component schematic*
The physical scheme of the c-PJU model can be seen in Figure 4, where the NodeMCU is installed in the breadboard along with other components. Each component and the NodeMCU are connected with a copper cable adjusted to its configuration pin.

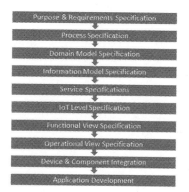

Figure 2.   Internet of Things design methodology.

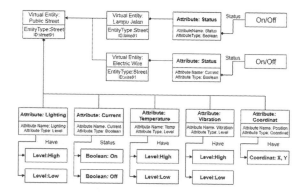

Figure 3.   Information model specification from e-PJU.

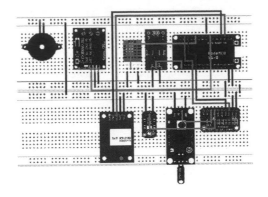

Figure 4.   Breadboard scheme of the e-PJU system

4   CONCLUSION

From the research that has been done, we can conclude the following:

1. Internet of Things technology can be implemented along with conventional technology in order to increase the capabilities and security of the devices.
2. With the implementation of IoT technology, the manually operated main highway lighting devices become automatic.
3. To provide better security, the device is equipped with a Global Positioning System (GPS), vibration sensor, and buzzer.

This research has benefits for those who wish to build an IoT device that can control lighting electronically. The model is ready for implementation by the government, and it could minimalize the monthly cost of the lighting. Compared to previous research, the model in this research is equipped with a GPS module in order to detect the current location of the device. Also, the existing software development methodology is not suitable for the use of IoT development. Thus, the current methodology used in this research can be an inspiration for IoT developers.

REFERENCES

Abdelaziz, A., Elhoseny, M., Salama, A. S., Riad, A. M., & Hassanien, A. E. 2018. Intelligent algorithms for optimal selection of virtual machines in a cloud environment: Towards enhanced healthcare services. *Advances in Intelligent Systems and Computing.*

Asghar, M. H., Negi, A., & Mohammadzadeh, N. 2015. Principle application and vision in Internet of Things (IoT). *International Conference on Computing, Communication and Automation*, 2015.

Bari, N., Mani, G., & Berkovich, S. 2013. Internet of Things as a methodological concept. *Proceedings of the 2013 4th International Conference on Computing for Geospatial Research and Application.*

Hu, C. L., Chan, T. K., Wen, Y. C., Tantidham, T., Sanghlao, S., Yimwadsana, B., & Mongkolwat, P. 2018. Lot-based LED lighting control in smart home. *IEEE Sensors Journal.*

Mahajan, S. A., & Markande, S. D. 2017. Design of intelligent system for indoor lighting. *Proceedings of the 2nd International Conference on Computing, Communication, Control and Automation*, Iccubea 2016.

Shahid, N., & Aneja, S. 2017 Internet of Things: Vision, application areas and research challenges. *Proceedings of the International Conference on Lot in Social, Mobile, Analytics and Cloud.*

Shahzad, G., Yang, H., Ahmad, A. W., & Lee, C. 2016. Energy-efficient intelligent street lighting system using traffic-adaptive control. *IEEE Sensors Journal.*

Engineering, Information and Agricultural Technology in the
Global Digital Revolution – Hendrawan & Wijayanti Dual Arifin (eds)
© 2020 Taylor & Francis Group, London, ISBN 978-0-367-33832-9

Estimating the water infiltration capability of weathered volcanic and alluvial soils based on their moisture content: A case study in Tembalang District, Semarang City, Indonesia

B. Sudarmanto
Universitas Semarang, Semarang, Indonesia

A. Gaber
Port Said University, Qism El-Zohour, Port Said Governorate, Egypt

ABSTRACT: Knowledge of soil infiltration capability is very important for planning to mitigate flooding hazards. Research into handling surface drainage and the conservation of excess water in the soil needs to be encouraged and continuously developed. The aim of this research was to figure out the relationship between infiltration rate and surface parameters, which are easily obtained from radar images from space. This information is related to the infiltration capacity, the soil saturation, and the correlation between saturation and the water table. Such an infiltration study was carried out in the eastern part of Semarang City, namely the Tembalang District, which is represented by alluvium and weathered volcanic sediments. Three infiltration test points were placed: two in volcanic territories and one in alluvium deposits. The method used was the measurement of single-ring infiltration, which emphasized the measurement of the attainment of constant infiltration and a change of water content during infiltration observations. The results revealed a negative linear correlation between the constant infiltration's time and changes in soil water content in volcanic soil, and a positive linear correlation in alluvium soil. This finding needs to be further tested for soil types in several different locations. Additional tests to examine the variable existence of a groundwater table found linkages between changes in water content, constant infiltration time, and the presence of a groundwater table.

Keywords: Constant Infiltration, Alluvium Soil, Volcanic Soil

1 BACKGROUND

Floods comprise the natural risks that cause the most damaging impacts in Indonesia. According to EM-DAT records quoted from Hadihardaja (2015), floods cause more than 50% of the total natural disasters that occur in Indonesia.

On Java, Indonesia's most populated island, flooding has become an annual phenomenon that routinely occurs in big cities, especially in cities located on the north coast of Java. The main cause of the flood disasters is low soil infiltration capability, causing the soil to no longer absorb water, abnormal rainfall, temperature changes, dam dikes, rapid melting of snow, and obstruction of water flow elsewhere (Ligal, 2008, cited in Sezar, 2017). Floods are also caused by sea level fluctuations, especially along alluvial coastal plains, swamp areas, and river encounters with alluvial plains (Dibyosaputro, 1984, cited in Adji, Nurjani, & Wicaksono, 2017). For tropical regions such as Indonesia, five important factors cause floods: rainfall, destruction of the watershed, river plumbing errors, river dredging, and regional faults and facility and infrastructure development (Maryono, 2005, cited in Sezar, 2017).

Thus it can be concluded that the root cause of all of these problems is limited absorption of surface deposits and the limited infiltration rate of surface water into the soil as part of the groundwater conservation cycle. This has become the consideration of the spatial policy-holder in determining the "zero delta Q policy" – a policy to minimize the impact of land use change – which results in the decrease of absorption of surface sediments or the increase of runoff through construction technology that compensates for the decreasing absorption of surface water with the replacement of the recharge function.

Researchers in the field of civil engineering have conducted a study on the innovation of water infiltration construction into artificial soil, which is currently quite popular and is believed to increase the absorption of the surface sediments, along with the construction of infiltration wells. However, the infiltration wells are only applicable in areas where the water table is below the depth of the infiltration well itself. Thus drainage system implementation that prioritizes the infiltration of excess runoff to the ground requires data based on reliably mapping the water table depth and that is easily accessible to planners and policy-holders.

2 RESEARCH PURPOSE

This research was intended to formulate the relationship between the different variables of soil that can be obtained by remote sensing such as air temperature, humidity, and soil moisture with the infiltration rate, including infiltration capacity and the time to reach constant infiltration.

This research was conducted on the basis of consideration of benefits, namely the utilization of water table maps for the purposes of marking groundwater conservation areas with the placement of infiltration and injection wells and retention ponds. For local governments, this research is supportive in technical terms, such as the realization of the "zero delta Q policy" in the sense of mathematical calculation of the number and dimensions of recharge wells in the scale of the area affected by land use change.

3 PROBLEM LIMITATION AND LOCATION

The surficial geology of Semarang consists of a coastal plain (Qa), river and lake deposits composed of clay, sand, and 50 meters or more of tuffaceous sandstone (Qtd), conglomerate, volcanic breccia (Qpkg), lava flows, tuffa, tuffaceous sandstone and claystone, massive marls (Tmpk) in the upper part, locally carbonaceous marls intercalating with tuffaceous sandstone, and limestone modules that are 3–200 centimeters in diameter. Few of the aforementioned surficial geological units have proper porosity and permeability and show significant potential for recharging rainfall. Figures 1 and 2 depict geologic maps of Semarang and its surroundings. Tembalang District has four soil typologies, namely Qa, Qtd, Qpkg, and Tmpk (Figure 1).

Figure 1.   Geological map of Tembalang sheets, central Java (Geological Research and Development Centre, Bandung, Indonesia, 1996)

Figure 2. The single-ring infiltrometer preparation and measurement at the Kramas and Sendangmulyo experiments.

4 METHODOLOGY

Efforts to find the water infiltration formulas were the first step in mapping the shallow groundwater potential zones. The Horton formula was used in this study as a reference. The research implementation is explained as follows:

1. Use a single-ring infiltrometer.
2. Determine the best represented locations with different properties of soils (soil type data in accompanying maps).
3. Measure infiltration following steps (source: Soil Research Institute, 2005).
4. Install a single-ring infiltrometer (depth = 30 cm, diameter = 30 cm) at the observation point.
5. Press with a bat (place the wood on the ring), ring 20 cm into the ground.
6. Install one sheet of plastic in a small ring to prevent damage to the soil during filling.
7. Record air temperature and humidity with a hygrometer.
8. Record the water content of the surface with a five-point test surrounding the ring (maximum in radius of 1 m).
9. Fill the infiltration ring with water carefully.
10. Start the measurement by pulling out the plastic sheet from the ring and running the stopwatch.
11. Record the initial water level by looking at the scale and recording the water drop in a certain time interval; the time interval depends on the speed of the water drop.
12. Add water if the water level is 2 cm from the ground, and record the height of the initial water level. Repeat until there is a constant decrease in water at the same time.
13. Stop testing after infiltration is constant.
14. Record the water content after infiltration measurements stop by using a moisture content measurement tool.

In the study, two parameters of field observation were selected, namely time to achieve constant infiltration and delta of surface water content. In the Horton formula ($f = fc + (fo - fc) e - Kt$), these two parameters are important in determining infiltration capacity. Time to achieve constant infiltration is represented by the variables fo and fc in the formula. The delta of surface water content parameter is represented by variable K, which is a constant whose amount depends on the soil's condition. The intended soil conditions then become the hypothesis of this study, and they can be summarized by changes in surface water content. Data records from this field observation were then analyzed to find relationship patterns in graphical form. The graph of the pattern of relations between the two parameters can be used to strengthen other soil condition variables in determining K, such as land cover, land use change, etc.

5 RESULTS

The infiltration experiment in Tembalang was carried out in two locations, Kramas and Sendangmulyo. In the Kramas subdistrict, the testing points were carried out at the Kramas Village Office, represented by the geological unit of Qpkg, and in the Sendangmulyo subdistrict there were two locations, namely Sendangmulyo 1, represented by the geological unit of Qa, and Sendangmulyo 2, represented by the geological unit of Qtd.

The Qpkg unit is the formation of volcanic breccia, lava flows, tuffs, tuffaceous sandstones, and clay stones. The flow and lava breccia formed with inserts of lava and fine to coarse coals. Local soil found below the claystone contained mollusks and tuff sandstone. Volcanic rocks, were also found, which were decaying, reddish brown, and often formed large blocks with thicknesses ranging from 50 to 200 meters.

The Qa unit represents coastal plains, rivers, and lake deposits. Coastal plains generally consist of clay and sand reaching a thickness of 50 meters or more. Sand deposits generally come from water-bearing layers of delta deposits as thick as 80 meters. River and lake deposits consist of gravel, sand, and silt 1 to 3 meters thick and lumps composed of andesite, limestone, and a little sandstone.

The Qtd unit refers to the Damar Formation of tuff sandstone, conglomerate, and volcanic breccia. Sandstone contains mafic, feldspar, and quartz minerals. Volcanic breccia may be deposited as lava. This formation is partly nonmarine; local mollusks are found and remnants of vertebrates are revealed around the Damar River. The results are represented by the following infiltration graph (Figure 3).

The testing points in Sendangmulyo 1 (Qa) and Kramas (Qpkg) resulted in an infiltration graph with as many as five observations.

Furthermore, at the observation point at Sendangmulyo (Qtd), there were obstacles in the observation because of heavy rain and long periods of time in achieving constant infiltration conditions (more than 10 hours of observation). For this reason, a special study will be carried out for this observation point during the dry season.

6 DISCUSSION

The infiltration experiments at three sites with three different geological units can be grouped into two patterns. The first is the pattern of infiltration in Qa representing alluvial soil. Time to achieve constant infiltration in Qa had a positive linear correlation with changes in water content (D water content = posttest – pretest) on the surface (Figure 4).

The second pattern is the pattern of infiltration in Qpkg, which represents the type of volcanic soil. Time to achieve constant infiltration in the Qpkg unit had a negative linear correlation with changes in water content on the surface (Figure 5).

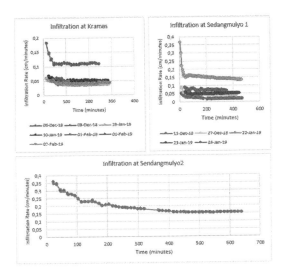

Figure 3. The infiltration graph of the Kramas and Sendangmulyo experiments.

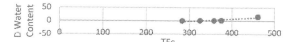

Figure 4. The infiltration pattern representing the alluvial soil.

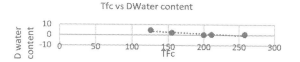

Figure 5. The infiltration pattern representing the volcanic soil.

For the results of the test in Qtd, we cannot conclude the relationship between the failed infiltration experiments.

The pattern of the relationship between TFc and delta water content as shown in Figures 4 and 5 has not produced an ideal graph from which conclusions. This is due to a lack of field observations that only amount to 5 times. Thus statistical validation has not been done in measuring the quality of the data. But at least, from the two graphs presented, there are significant differences between the two types of land with different geological types.

7 CONCLUSION AND RECOMMENDATION

The result of estimating the water infiltration rate in volcanic breccia and alluvial soils in relation with their water content has revealed a relationship between TFc and Delta water content that forms a different relationship pattern. This reinforces calculating soil infiltration capacity by taking into account the type of soil, especially the type of soil seen from its geological type. Changes in surface water content that can be analyzed by processing satellite-based map data (optical or radar) cannot be generalized to all land types.

This experiment encourages further research to conduct additional experiments on selected locations after mapping the shallow groundwater table. This is done because the infiltration of water into the soil is influenced by several factors besides the natural moisture content of the soil and the type of soil: the depth to the shallow unconfined groundwater table, the porosity and permeability of the soil, and the land slope and type of land cover will all be considered in the next phase of the project.

REFERENCES

Adji, T. N., Nurjani, E., & Wicaksono, D. 2017 Identifikasi potensi airtanah pada area dengan beragam bentuklahan menggunakan beberapa parameter lapangan dan pendekatan SIG di kawasan parangtritis, DIY. INA-Rxiv.

Directorate of Water Management. 2007. General guidelines for the establishment of resilient wells in the framework of anticipating drought in 2007.

Hadihardaja. 2015. Zero delta Q policy for managing flood risk. Conference Paper, Semarang, Indonesia.

National Standardization Agency. 2000. Procedures for planning of rainwater infiltration wells for landscapes.

Sezar, R. 2017. Definisi Banjir Menurut Para Ahli. Uploaded on Scribd.

Sunjoto, 1988, in Suripin. 2000. Soil and water conservation. Master of Civil Engineering, Diponegoro University, Semarang.

Sunjoto. 2011. Pro-air drainage technique. Drainage Engineering, Department of Civil and Environmental Engineering, Gadjah Mada University, Yogyakarta, p. 41.

Susilo et al. 2017. Kajian empiris sumur resapan pada tanah silt (Empirical study of recharge wells on wilt). http://journals.usm.ac.id/index.php/teknika/issue/view/64

Tauro, F. et al. 2018. Measurements and observations in the XXI century (MOXXI): Innovation and multidisciplinarity to sense the hydrological cycle. Hydrological Sciences Journal. www.tandfonline.com/loi/thsj20.

USDA. 2017. www.nrcs.usda.gov/wps/portal/nrcs/detail/soils/survey/

Undang Kurnia et al. 2010. The retention of groundwater. http://balittanah.litbang.pertanian.go.id

*Engineering, Information and Agricultural Technology in the
Global Digital Revolution – Hendrawan & Wijayanti Dual Arifin (eds)*
© 2020 Taylor & Francis Group, London, ISBN 978-0-367-33832-9

Feature selection for optimizing the Naive Bayes algorithm

Titin Winarti & Vensy Vydia
Universitas Semarang, Semarang, Indonesia

ABSTRACT: Naive Bayes is a data-mining method used in the classification of text-based documents. The advantage of this method is simple algorithms with low calculation complexity. However, Naive Bayes has a weakness where the independence of the Naive Bayes feature cannot always be applied so that it will affect the accuracy of calculations. Naive Bayes therefore needs to be optimized by giving scale using a gain ratio. Weighting with Naive Bayes raises problems in calculating the probability of each document, where many features that do not represent the tested class appear so that there is a misclassification. so weighting with Naive Bayes is still not optimal. This article proposes the optimization of Naive Bayes through using the weighting gain ratio, which is a method of selecting features in the case of text classification. The results of this study indicated that the Naive Bayes optimization method using feature selection and weighting gain ratio produces an accuracy of 94%.

1 INTRODUCTION

Classification is the process of sorting objects into a class, group, or category based on predetermined procedures, characteristics, and definitions (Gu et al., 2017). One form of classification is the classification of documents or text. Document or text classification is a field of research in information processing. The purpose of document classification is to develop a method for automatically determining or categorizing a document into one or more groups based on its content (Lytvyn et al., 2017). In this era, the grouping of text or documents is used when searching for a document. Therefore, quick and easy grouping of documents is very important.

Grouping documents is carried out by labeling the document category. This requires considerable time in classifying documents. A method is needed that can be used in the process of classifying or grouping documents quickly and accurately.

One classification method commonly used is Naive Bayes. Naive Bayes is a very simple and effective approach to classification learning (Granik & Mesyura, 2017). According to An, Sun, and Wang (2017), Naive Bayes is the possibility of labeling data classes or class attributes.

Naive Bayes has several advantages, namely simple algorithms, fast calculation, and high accuracy (Zhang et al., 2016). However, Naive Bayes also has a weakness in that a probability cannot measure the accuracy of a prediction. Therefore, Naive Bayes needs to be optimized by giving weight using the gain ratio. Weighting to Naive Bayes raises problems in calculating the probability of each document. There is a misclassification because the features that do not represent the tested class appear many times. Therefore, the weighting of Naive Bayes is still suboptimal.

2 METHODS

Naive Bayes is an algorithm that is effective and efficient in the classification process (Jiang et al., 2016). Figure 1 shows the weighted Naive Bayes proposed method using the gain ratio.

2.1 *Dataset*

The data used in this study were taken from online media: Kompas, Detik, and Tempo. Then came the process of determining basic words, general words that often appear or stopwords, and categories. The data processing is depicted in Figure 2.

2.2 *Preprocessing*

Preprocessing is the initial classification of documents that aims to prepare data to be structured. The results of preprocessing will be a numerical value so that they can be used as a source of data that can be further processed. Preprocessing is divided into several processes consisting of case folding, tokenizing, filtering, stemming, and calculation of word weighting.

Figure 2 shows the preprocessing process. Case folding is the initial stage of preprocessing text that converts text characters into all lowercase letters (Silge & Robinson, 2017). Accepted characters are only a to z. Non-letter characters are removed and considered a delimiter. Tokenizing is the stage of cutting the string input based on each word that composes it (Silge & Robinson, 2017). Filtering is the process of determining what words will be used to represent documents. In addition to describing the contents of the document, this term is also useful for distinguishing between documents. This process is done by taking important words from the token and removing stopwords. Stopwords are words that are not descriptive so they can be discarded or omitted and have no effect on the process (Silge & Robinson, 2017). In Indonesian, examples of stopwords include *yang, dan, dari, di, seperti,* and others. Stemming is the stage of finding the root (*akar*) word in the word filtering results. At this stage, the process of taking various forms of words is carried out with the same representation. The stem (root word) is a part of a word that remains after it has been eliminated (prefix and suffix). An example of a stem is *beri,* from "to give," "is given," "giving," and "gift."

2.3 *Weight calculation*

Naive Bayes is used in statistics to calculate the potential of a hypothesis. Naive Bayes calculates the potential of a class based on the attributes that are owned and determines which class

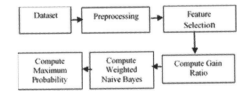

Figure 1. Experimental flow.

Figure 2. Preparation data set.

has the highest probability. Naive Bayes classifies classes based on simple probabilities by assuming that each attribute in the data is mutually exclusive. Naive Bayes is widely used based on some simple characteristics; Naive Bayes classifies data based on the probability of the P attribute x of each data class y (Jiang et al., 2016).

The Naive Bayes calculation is the probability of the emergence of "xa" document in the "yk" class category P (xa | yk), multiplied by the probability of the category P class (yk). From the results, division is then performed on the probability of occurrence of the document P (xa) (Jiang et al., 2016). The optimal class selection process is then carried out so that the greatest opportunity value of each class probability is chosen (Jiang et al., 2016).

Weighting class attributes can increase the influence of predictions (Kharya & Soni, 2016). By calculating the weight of attributes to the class, the basis for the accuracy of classification is not only the probability but also the weight of each attribute to the class. Weighted Naive Bayes is calculated by adding "wi" weights to each attribute.

Weighting can be formulated using the gain ratio. Each attribute of the gain ratio is multiplied by the number of "n" data, then divided by the average gain ratio of all of the attributes (Kharya & Soni, 2016).

The attribute of the gain ratio itself is the result of mutual information and entropy. Mutual information (MI) is a measuring value that states the attachment or dependence between two or more variables. The general measuring unit is used to calculate MI is bits, so it uses base logarithms (log) 2. Formally, MI is used between two variables; they are A and B as defined by Kulback and Leibler. Besides MI, entropy is used as a divider from MI in order to determine which attributes are the best or the most optimal.

Before getting the gain ratio value, entropy (E) is searched for. Entropy is used to determine how informative an input attribute is in order to produce an attribute output. Entropy calculation is completed by summing probabilities (Kharya & Soni, 2016). Therefore, the calculation of the gain ratio is the result of the calculation of MI divided by the results of the entropy calculation. The weighted calculation process of Naive Bayes using the gain ratio is divided into two stages. The first stage is the training process. In the training process, training data are taken then preprocessing is done. After that, the probability of per-category terms and the opportunity category (class) are calculated. The second stage is the process of testing. In the testing process, the test data are taken, then preprocessing is performed. Next, the value of the gain ratio is taken for each word and category.

Afterward, the "R" word is carried out by the ranking process (the number of words specified). From as many words as "R," the gain ratio is calculated. In the evaluation phase, the aim is to determine the accuracy of the results of the weighted Naive Bayes method. In the testing process, this is known as the Confusion Matrix, which represents the truth of a classification (Deng et al., 2016).

3   RESULTS

Testing of the results using the weighted Naive Bayes method is executed by comparing the results of the Naive Bayes experiment without using weighting. Comparisons were made to news documents of 65 documents in trial 1 and 145 documents in trial 2. The results compared were the accuracy of the data produced by calculating the difference between the weighted Naive Bayes and Naive Bayes. Next trial 1 of the proposed method used training data from as many as 35 documents and 30 test documents.

In trial 2, the test data used comprised 110 documents and the training data used were the same as those used in trial 1. The training data consisted of seven categories: football, automotive, health, technology, economy, politics, and law. Each category contained five documents. From the results of trial 1, the results of the Naive Bayes accuracy were found to be 92%, while those of the weighted Naive Bayes were 94%. In addition, from the results of trial 2 were obtained a Naive Bayes accuracy of 92% and a weighted Naive Bayes accuracy of 84%. The results can be seen in Table 1.

Table 1. Accuracy result.

Method	Accuracy %	
	Trial 1	Trial 2
Naive Bayes	92	92
Weighted Naive Bayes	94	84

Table 2. Feature selection.

The Best Term	Proposed Method %	Naive Bayes %
50	91	95
30	91	95
10	94	91

Based on trial 2, the selection process for R (50, 30, and 10 best terms) was carried out. From the results of the selection of features using the best 50 and 30 terms, an accuracy of 91% was obtained, but in trial 2, the accuracy of the proposed method tended to be low compared to ordinary Naive Bayes. This is because the terms that often appear in all document categories result in a high gain ratio and misclassification. The accuracy results in trial 2 were low. So the process of selecting the best features needs to overcome classification errors caused by the frequent appearance of all documents. From the results of the trial selection of features using the best 50 and 30 terms, an accuracy was obtained of 91% for the proposed method and 95% for Naive Bayes. When using the best 10 terms, the accuracy was 94% for the proposed method and 91% for the ordinary Naive Bayes method. The results of the trial on feature selection can be seen in Table 2.

4 DISCUSSION

From the results of trial 1, the Naive Bayes accuracy value was 92% while the accuracy value for the proposed weighted Naive Bayes was 94%. The results of the proposed method are higher due to the weighting of the probability of each word in the document against the category. Giving weights on probabilities evokes the distance between one-word opportunities and the category. The results of the proposed study are in accordance with the research of Kharya and Soni (2016), who conclude that weighting class attributes can improve predictive outcomes.

In trial 2, the accuracy of the proposed method tended to be low compared to ordinary Naive Bayes. This is because the terms that often appear in all document categories result in a high gain ratio and misclassification. The accuracy results in trial 2 were low. So the process of selecting the best features to overcome classification errors caused the frequent appearance of all documents. From the results of the trial selection of features using the best 50 and 30 terms, an accuracy of 91% was obtained for the proposed method and 95% for Naive Bayes. This is because the terms that often appear in other classes are also found in the class being tested. When using the best 10 terms, the accuracy was 94% for the proposed method and 91% for the ordinary Naive Bayes. This is because the term used in the class being tested represents the class. So in this trial, it is known that the selection of the best features can reduce the number of terms that often appear in other classes.

5 CONCLUSIONS

The weighted Naive Bayes method can optimize the accuracy of the ordinary Naive Bayes method. This can be seen from the results of weighted Naive Bayes of 94% accuracy compared to an ordinary Naive Bayes accuracy of 92%. The contribution of this article is to show that weighted Naive Bayes can produce a higher level of accuracy because each probability of an attribute is given a weight, which results in a higher value. When selecting the best features using 10 terms, the accuracy was 94% for the proposed method and 91% for the normal Naive Bayes method. It can be concluded that the selection of features can overcome classification errors.

REFERENCES

An, Y., Sun, S., & Wang, S. 2017. Naive Bayes classifiers for music emotion classification based on lyrics. *2017 IEEE/ACIS 16th International Conference on Computer and Information Science (ICIS)*, 635–638.

Deng, X. et al. 2016. An improved method to construct basic probability assignment based on the confusion matrix for classification problem. *Information Sciences, 340*, 250–261.

Granik, M., & Mesyura, V. 2017. Fake news detection using naive Bayes classifier. *2017 IEEE First Ukraine Conference on Electrical and Computer Engineering (UKRCON)*, 900–903.

Gu, H. et al. 2017. An object-based semantic classification method for high resolution remote sensing imagery using ontology. *Remote Sensing, 9*(4), 329.

Jiang, L. et al. 2016. Deep feature weighting for naive Bayes and its application to text classification. *Engineering Applications of Artificial Intelligence, 52*, 26–39.

Kharya, S., & Soni, S. 2016. Weighted naive bayes classifier: A predictive model for breast cancer detection. *International Journal of Computer Applications, 133*(9), 32–37.

Lytvyn, V. et al. 2017. Classification methods of text documents using an ontology based approach. In *Advances in Intelligent Systems and Computing* (pp. 229–240). New York: Springer.

Silge, J., & Robinson, D. 2017. *Text Mining with R: A Tidy Approach*. Sebastopol, CA: O'Reilly Media, Inc.

Zhang, L. et al. 2016. Two feature weighting approaches for naive Bayes text classifiers. *Knowledge-Based Systems, 100*, pp. 137–144.

Engineering, Information and Agricultural Technology in the Global Digital Revolution – Hendrawan & Wijayanti Dual Arifin (eds)
© 2020 Taylor & Francis Group, London, ISBN 978-0-367-33832-9

Laboratory analysis of the effect of Sulfate in tidal water on the performance of asphalt mixture

B.H. Setiadji, S.P.R. Wardani & K.P. Azizah
Universitas Diponegoro, Semarang, Indonesia

ABSTRACT: It is known that Sulfate is one of chemical elements in tidal water that may damage submerged roads pavement, however, how much the impact required further analysis. This study aims to evaluate the durability of asphalt mixture against tidal water by conducting a simulation which involves different tidal water sources and Sulfate solutions. To evaluate the durability of the mixtures, durability indices were used. The results showed that the amount of Sulfate in tidal water has a negative impact on the mixture performance. The use of Starbit E-55 polymer asphalt can reduce the negative impact of tidal water up to 45%, although the level of the reduction in strength loss from the use of this kind of asphalt may vary depending on the location of tidal water, amount of decayed-organic/non-organic materials in tidal water, and so on.

1 INTRODUCTION

Negative impact of environmental factors on pavement structure is still an interesting research topic to date. People are trying to find out not only about how these environmental factors affect the degradation of pavement structure performance, but also about the magnitude of the damage and the most suitable solutions that can be proposed to overcome the impact (Xiong *et al.*, 2011; Ahmad *et al.*, 2014). Related to these environmental influences, tidal water is actually a phenomenon that is common in coastal areas. Due to poor drainage channels and low topographical conditions, this water may also inundate road pavement. This condition is also encountered in some cities at northern Java corridor where are located near the coastline, such as Semarang City. At rainy season, the rising sea level will contribute on the tidal flooding and inundate asphalt pavement road construction.

The submerged road pavements have a tendency to be repeatedly damaged, even though regular repairs have been made. Why does this tidal water have a very large damage on road pavement? To answer this question, Setiadji and Utomo (2017) stated that tidal water has a higher level of damage compared to ordinary water, however, the level of damage given by different tidal water on road pavement can varies, which may be caused by two factors: (i) the chemical content contained by waste carried by tidal water through a region; and (ii) high-content of single chemical compound or combination of chemical compounds and interaction between them. The use of high-content Chloride Setiadji and Utomo (2017) has given the required evidence for the hypothesis in point (ii) above, however, it also need to be proved that other chemicals that often found in tidal water, such as Sulfate, can also provide similar effects.

Therefore, the purpose of this study is to simulate the impact of Sulfate in tidal water on the durability of road pavement. Besides, it was also examined the use of modified asphalt in the mixture to increase the road pavement durability against tidal water. All research work was carried out in the laboratory, while the implementation may be carried out at a later stage.

2 RESEARCH METHODOLOGY

The research methodology consists of several parts as follows:

a. Material property tests and mixture preparation. The type of asphalt mixture used in this study is asphalt mixture wearing course (AC-WC). This mixture was composed of dense-graded aggregate and asphalt Penetration 60/70. As an alternative, the use of Starbit E-55 Polymer Asphalt was proposed. All material properties have to be tested to ensure that they could meet the requirements issued by Directorate General of Highway (2018).
b. Mixture treatment. To measure the durability of the mixtures, they were treated by immersing them in tidal water collected from two locations in Semarang City, i.e. Terboyo and PRPP, and sulfate solutions for different soaked periods: 1, 3 and 7 days.
c. Analysis. To examine the durability of the mixtures under different treatment, two durability indices (Craus, Ishai and Sides, 1981) was used. The first index (r) is defined as the number of sequential slopes of the durability curve. This index can be calculated using the following equation:

$$r = \sum_{i=0}^{n-1} \frac{S_i - S_{i+1}}{t_{i+1} - t_1} \tag{1}$$

in which: S0 = absolute value of initial strength (kg); S_i, $S_i + 1$ = remaining strength at time ti and ti+1 (kg); ti, ti + 1 = immersion period (starting from the beginning of the test)

The second durability index (a) is defined as the area of losing strength of one day. A positive value of a indicates a loss of strength, while a negative one is an increase in strength. By definition, a <100. Therefore, it is possible to state the percentage of the remaining strength of one day (Sa), as shown by the following equation:

$$S_a = (100 - a) \tag{2}$$

3 RESULTS AND ANALYSIS

3.1 *Material and tidal water properties*

The material used is local aggregate of three fractions, i.e. ¾ and ½ in. maximum-size coarse aggregate fraction and fine aggregate fraction, and two types of asphalt, i.e. Pertamina Pen 60/70 asphalt and Starbit E-55 polymer asphalt. In this study, the Starbit E-55 asphalt was used to observe whether this asphalt can improve the resistance of the mixture against chemicals or not. To ensure that the properties of aggregate and asphalt meet the specifications, it is necessary to conduct several tests on those materials. The material tests conducted in this study showed that all the properties can meet the requirements.

Four kinds of water were proposed in this research, they are: distilled water, sea water, two tidal waters from different locations, and sulfate solutions. A chemical test was carried out to determine the type and amount of chemical content contained in the waters. Six parameters were selected (as seen in Table 1) to represent the characteristic of each water.

To analyze the characteristics of the two tidal waters in Table 1, sea water and distilled water were used as references. As seen in the table, the amount of Sulfate and Chloride are increasing from left (distilled water) to right (sea water). The amount of them, together with the amount of bi-carbonate alkalinity (in the opposite trend), will affect on the value of pH.

The table also shows that tidal water from different location will consist of different chemical compounds, although they come from similar sea water. Setiadji and Utomo (2017) stated that the types of decayed-organic or non-organic materials/wastes that mixed with the sea water at the time the sea water overflow to mainland was one factor that could determine the chemical

Table 1. Results of chemical tests of water from different sources.

No	Parameter	Source of water			
		Distilled water	Tidal Water at Terboyo	Tidal Water at PRPP	Sea water
1	Salinity (%)	0.03	3.48	8.58	2.94
2	pH	6.7	7.5	7.6	8.3
3	Sulfate (mg/l)	< 3.8	96.6	419	1593
4	Chloride (mg/l)	15	1824	4674	18244
5	Alkalinity CO3 (mg/l)	0	0	0	0
6	Alkalinity HCO3 (mg/l)	195	471	354	259

compounds that composing the tidal water. In addition, the change of the chemical compounds in tidal water could also be contributed by the mixture between sea water and surface water (rainwater, sewage water, and so on). From the table, it is very clear to see that tidal water at PRPP could produce more damage on road pavement than that tidal water at Terboyo.

To simulate the effect of Sulfate on the mixture performance, three Sulfate solutions were proposed, that is, 5000 mg/l and 2000 mg/l, to resemble the amount of sulfate in tidal water at PRPP and Terboyo, respectively, and additional amount of 500 mg/l Sulfate solution.

3.2 Durability of the mixtures

Durability of the mixture is measured based on three immersion periods. By measuring durability of the mixtures, ones could determine in which period the mixture gains or loses the highest strength and so on. As mentioned above, in this research the mixtures experience three immersion periods: 24, 72 and 168 hours (1, 7 and 14 days). The performance of the mixtures then was assessed by using parameters percentage of remaining strength (Sa) and the second durability index (a), as shown in Figure 1 and Table 2, respectively.

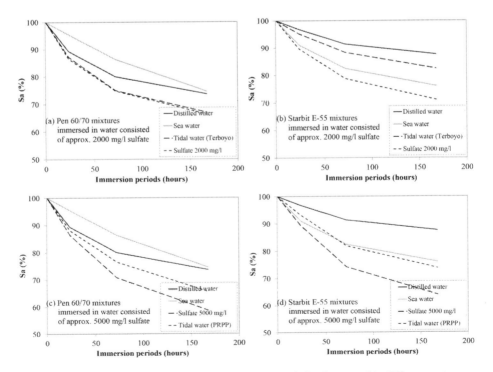

Figure 1. Percentage of remaining strength of mixtures after being immersed in different water.

Table 2. The second durability index (a) of mixtures after being immersed in different waters.

Type of asphalt	Immersion periods (hours)	Source of water						
		Distilled water	Sea water	Tidal water at Terboyo	Tidal water at PRPP	500 mg/l Sulfate solution	2000 mg/l Sulfate solution	5000 mg/l Sulfate solution
Pen 60/70	0.5	0.00	0.00	0.00	0.00	0.00	0.00	0.00
	24	10.61	4.52	12.73	11.90	12.22	13.38	13.57
	72	9.32	9.06	12.23	11.49	10.48	11.78	15.56
	168	6.41	11.82	8.06	11.01	7.41	8.57	12.23
	Average	8.78	8.46	11.01	11.47	10.03	11.24	13.79
Starbit E-55	0.5	0.00	0.00	0.00	0.00	0.00	0.00	0.00
	24	3.17	8.90	4.80	6.65	11.42	10.29	10.48
	72	5.43	8.59	6.84	11.39	9.99	10.89	15.28
	168	3.73	6.40	5.90	8.15	7.06	7.74	10.30
	Average	4.11	7.96	5.84	8.73	9.49	9.64	12.02

In Figure 1, it can be seen that the remaining strength of mixtures with Starbit E-55 after immersion was higher than that of mixtures with Pen 60/70. From Figure 1, it can be concluded that the reduction in strength is not the same between tidal water and Sulfate solution even though both have almost the same Sulfate content. Generally, the mixture immersed in Sulfate solution will experience a more significant loss of strength compared to tidal water. The mixture of two or more chemical compounds in tidal water may reduce the destructive ability compared to that of Sulfate when used as a single chemical compound. This was not found in Setiadji and Utomo (2017) where Chloride as a single chemical compound in a Chloride solution has smaller destructive ability than that of using tidal water. This does not mean that Sulfate has a higher destructive ability than Chloride. To arrive at this conclusion, a further research is needed.

Table 2 shows that the highest strength loss generally occurs at the first period (after 24 hours), although some of them experienced the highest strength reduction in the second or third period of immersion (Setiadji, Wardani and Perdana, 2015). The latter can occur because the setting process of achieving optimum strength from the mixture has occurred in the first period of immersion (Siswosoebrotho, Karsaman and Setiadji, 2003). The use of Starbit E-55, which contains elastomers which have chemical resistance properties, could improve the resistance of the mixtures against tidal water and Sulfate solution up to 45% & 25%, respectively.

4 CONCLUSIONS

This paper presented the evaluation of the impact of sulfate in tidal water and the use of modified asphalt to reduce this impact. The results showed that the amount of sulfate in tidal water has a negative impact on the mixture performance and the use of asphalt Pen 60/70 is insufficient for roads that are often submerged in tidal water. The use of Starbit E-55 polymer asphalt can reduce the negative impact of tidal water up to 45%, although the level of the reduction in strength loss from the use of this kind of asphalt can vary depending on the location of tidal water, amount of decayed-organic or non-organic materials/wastes in tidal water, and so on.

REFERENCES

Ahmad, J. et al. (2014) 'Investigation into hot-mix asphalt moisture-induced damage under tropical climatic conditions', Construction and Building Materials. Elsevier, 50, pp. 567–576.
Craus, J., Ishai, I. and Sides, A. (1981) 'Durability of bituminous paving mixtures as related to filler type and properties (with discussion)', in Association of Asphalt Paving Technologists Proceedings.

Directorate General of Highway (2018) *General Specification for Road and Bridge Construction*. Jakarta.

Setiadji, B. H. and Utomo, S. (2017) 'Effect of chemical compounds in tidal water on asphalt pavement mixture', *International Journal of Pavement Research and Technology*. Elsevier, 10(2), pp. 122–130.

Setiadji, B. H., Wardani, S. P. R. and Perdana, S. (2015) 'Durability of Road Pavement Against Tidal Inundation', *Journal of Society for Transportation and Traffic Studies*, 6(3), pp. 1–11.

Siswosoebrotho, B. I., Karsaman, R. H. and Setiadji, B. H. (2003) 'Development of a cyclic water vapour test for durability assessment of bituminous mixtures for pavement material', *Journal of the Eastern Asia Society for Transportation Studies*, 5(6), pp. 940–950.

Xiong, R. *et al.* (2011) 'Durability of asphalt mixture in sulfate corrosion environment', *Journal of Chang'an University (Natural Science Edition)*, 6.

*Engineering, Information and Agricultural Technology in the
Global Digital Revolution – Hendrawan & Wijayanti Dual Arifin (eds)
© 2020 Taylor & Francis Group, London, ISBN 978-0-367-33832-9*

Best employee selection in Central Java using the ELECTRE method

Saifur Rohman Cholil, Agusta Praba Ristadi Pinem, Aria Hendrawan, Titis Handayani
& Atmoko Nugroho
Universitas Semarang, Semarang, Indonesia

ABSTRACT: Employee selection can be said to be a valuable asset to a company because it contributes to and guarantees the development of the company. The purpose of the selection is to develop a decision support system (SPK) for selecting the best employees. This study used questionnaire data from the Department of Trade and Industry (Industry and Trade) Provincial Jawa T, puffing for the calculation process from quantitative to qualitative amended with the ELECTRE method. The criteria for research reference were service orientation (C1), integrity (C2), commitment (C3), discipline (C4), and collaboration (C5). Each criterion had sub-criteria. The selection of the best employees using ELECTRE can lead to better performance of the selected employees and motivate other employees. The employees with the greatest value are the employees who are selected as the best employees.

Keywords: ELECTRE, Linked Scale, Best Employee

1 INTRODUCTION

The choice of the best employee in a company will determine the success of the company itself. Employees who work hard and to the fullest in carrying out their work and responsibilities will make the company even better in the future. Selecting the best employees can also motivate the other employees (Fitriah & Irfiani, 2018). This research was the result of an interview with Mr. Noor Aziz, the head of the industry standardization section of the Department of Industry and Trade of Central Java Province (DISPERINDAG), concerning the selection of the best employees in an attempt to increase employee performance every year.

In 2012, at DERPERINDAG, the best employee selection was done manually. In this case, the author applied a decision support system (SPK) as one of the methods of decision-making in his research. The author used Elimination and Choice Expressing (ELECTRE) as the decision support system. The ELECTRE method was used because it can analyze wisely by involving qualitative and quantitative criteria (Putra, Andreswari, & Susilo, 2015). This method can also make assessments and rankings obtained by comparing pairs of each of the same criteria (Cholil & Indriyawati, 2019; Fauzi, 2016). The selection of the best employees certainly must meet the criteria determined by the head of the office and the questionnaire data of the employees approved by the section head. To simplify the calculation of results, a linked scale was utilized in this study. The application carried out uses five scales, and each scale has a value (Maryuliana, Subroto, & Haviana, 2016).

2 RESEARCH METHODS

The research method consisted of collecting data using questionnaires on each assessment criteria. The data obtained qualitatively were changed into quantitative form. The criteria for evaluating employee performance in the questionnaire were as follows:

C1 = Service Orientation
C2 = Integrity
C3 = Commitment
C4 = Discipline
C5 = Cooperation

The respondents were asked to weigh each assessment criteria: 20 (C1), 15 (C2), 10 (C3), 40 (C4), 15 (C5).

The calculation of results was carried out using a linked scale. The linked scale is a psychometric scale that is widely used in questionnaires (Helmi, Munjin, & Purnamasari, 2016). In this study, the linked scale was used to measure several questions (Budiaji, 2013) regarding the best employee assessment criteria. The choices in each question were:

a. Strongly disagree = 1
b. Disagree = 2
c. Doubt = 3
d. Agree = 4
e. Strongly agree = 5

3 RESULTS AND DISCUSSION

Decision support systems using ELECTRE rank the same comparison or alternatives according to criteria (Nashrullah Irfan, Abdillah, & Renaldi, 2015). Data from the questionnaire results were processed as follows.

Information: The alternative used in the study was 13 employees: A1 = Okta Riana, A2 = Perel Dwi, A3 = Vina Agustina, A4 = Salma Anindita, A5 = Anissa Hapsari, A6 = Hidayat Abdullah, A7 = Riski Aditya, A8 = Agung Pambudi, A9 = Kardiyono, A10 = Joko Susilo, A11 = Irfan Halim, A12 = Erika Rahayu, A13 = Noor Aziz.

The assessment criteria have sub-criteria applied in several questions in each criterion. Criterion C1 has five sub-criteria, C2 has six sub-criteria, C3 has three sub-criteria, C4 has eight sub-criteria, and C5 has six sub-criteria.

3.1 *Normalization of a matrix (R)*

The sub-criteria average value was then normalized:

Table 1. Sub-criteria value of each alternative.

ALTERNATIVE	C1					C2						C3			C4								C5					
	1	2	3	4	5	1	2	3	4	5	6	1	2	3	1	2	3	4	5	6	7	8	1	2	3	4	5	6
A1	5	4	4	4	4	5	4	4	5	4	5	4	4	4	5	2	4	4	4	4	4	4	4	4	4	4	4	4
A2	4	5	4	4	4	5	4	4	4	4	4	4	4	4	2	4	4	4	4	4	4	4	4	4	4	4	4	4
A3	5	4	4	4	4	5	4	5	4	4	5	4	5	4	4	4	4	5	4	4	4	4	4	4	4	4	4	4
A4	5	4	4	4	4	5	4	4	4	4	4	4	4	4	2	4	4	4	4	4	4	4	4	4	4	4	4	4
A5	5	5	5	5	5	5	5	5	5	5	5	5	5	5	1	5	5	5	5	5	4	5	5	5	5	4	5	5
A6	4	4	2	2	4	4	4	2	4	4	4	4	4	2	2	4	4	4	2	2	2	4	4	4	4	4	4	4
A7	4	4	4	4	4	4	4	2	4	4	4	4	4	4	2	4	4	4	4	4	2	4	4	4	4	4	4	4
A8	4	4	5	4	4	5	4	4	3	4	3	2	4	5	1	4	5	4	5	3	3	4	5	5	4	4	5	5
A9	4	5	4	4	4	5	4	2	4	4	5	2	3	4	4	1	4	4	4	4	4	4	4	3	4	5	3	4
A10	4	4	4	4	4	5	5	4	4	4	5	4	4	4	4	4	4	4	4	4	4	4	4	4	4	4	4	4
A11	4	5	4	4	4	4	4	4	4	4	4	4	4	5	1	4	4	4	4	4	4	3	4	4	4	4	4	4
A12	4	4	5	5	5	4	3	5	4	4	5	4	5	4	4	2	4	4	5	5	4	4	5	4	4	4	4	5
A13	4	4	4	4	4	5	4	4	5	4	5	4	4	5	4	2	4	4	5	4	3	3	4	4	4	4	4	5

$$r_{ij} = \frac{x_{ij}}{\sqrt{\sum_{i=1}^{m} x^2_{ij}}}$$

so that the R matrix was obtained:

$$R = \begin{bmatrix} 0,271 & 0,320 & 0,273 & 0,297 & 0,246 \\ 0,271 & 0,256 & 0,273 & 0,266 & 0,246 \\ 0,271 & 0,320 & 0,305 & 0,420 & 0,275 \\ 0,271 & 0,256 & 0,273 & 0,226 & 0,246 \\ 0,424 & 0,448 & 0,381 & 0,367 & 0,430 \\ 0,121 & 0,162 & 0,273 & 0,066 & 0,246 \\ 0,243 & 0,162 & 0,273 & 0,188 & 0,246 \\ 0,271 & 0,222 & 0,167 & 0,197 & 0,344 \\ 0,271 & 0,203 & 0,167 & 0,88 & 0,206 \\ 0,243 & 0,320 & 0,273 & 0,376 & 0,246 \\ 0,271 & 0,229 & 0,273 & 0,210 & 0,213 \\ 0,339 & 0,248 & 0,305 & 0,332 & 0,308 \\ 0,243 & 0,320 & 0,305 & 0,223 & 0,275 \end{bmatrix}$$

3.2 Normalized matrix weighting (V)

After normalization, each column of the matrix R was multiplied by the weight (Wj) in the equation: V = R * W while the weight value (W) of each criterion was 20 (C1), 15 (C2), 10 (C3), 40 (C4), and 15 (C5). The value of V was then obtained:

$$V = \begin{bmatrix} 5,422 & 4,807 & 2,726 & 11,885 & 3,690 \\ 5,422 & 3,846 & 2,726 & 10,630 & 3,690 \\ 5,422 & 4,807 & 3,048 & 16,807 & 4,126 \\ 5,422 & 3,846 & 2,726 & 10,630 & 3,690 \\ 8,473 & 6,718 & 3,810 & 14,681 & 6,447 \\ 2,425 & 2,432 & 2,726 & 2,657 & 3,690 \\ 4,850 & 2,432 & 2,726 & 7,517 & 3,690 \\ 5,422 & 3,030 & 1,670 & 7,878 & 5,158 \\ 5,422 & 3,040 & 1,670 & 7,517 & 3,095 \\ 4,850 & 4,807 & 2,726 & 15,033 & 3,690 \\ 5,422 & 3,440 & 2,726 & 8,404 & 3,196 \\ 6,778 & 3,724 & 3,048 & 13,287 & 4,613 \\ 4,850 & 4,807 & 3,048 & 8,913 & 4,126 \end{bmatrix},$$

3.3 Look for matrix concordance and discordance

1. Determining the value in the concordance matrix required adding the weights included in the concordance set:

$$A_{12} = w_1 + w_2 + w_3 + w_4 + w_5$$
$$= 20 + 15 + 10 + 40 + 10$$
$$= 100$$

Calculation C_{13} arrived at C_{1312} done in the same way, so the concordance matrix value was obtained:

$$C = \begin{bmatrix}
- & 100 & 35 & 100 & 0 & 100 & 100 & 100 & 100 & 60 & 100 & 15 & 75 \\
45 & - & 20 & 100 & 0 & 100 & 100 & 85 & 100 & 45 & 100 & 15 & 60 \\
100 & 100 & - & 100 & 0 & 80 & 100 & 85 & 100 & 100 & 100 & 65 & 100 \\
45 & 100 & 20 & - & 0 & 100 & 100 & 85 & 85 & 45 & 100 & 15 & 40 \\
100 & 100 & 100 & 100 & - & 100 & 100 & 100 & 100 & 60 & 100 & 100 & 100 \\
25 & 25 & 0 & 25 & 0 & - & 40 & 10 & 25 & 25 & 10 & 0 & 0 \\
25 & 45 & 0 & 25 & 0 & 100 & - & 30 & 65 & 45 & 25 & 0 & 20 \\
35 & 35 & 35 & 35 & 0 & 75 & 90 & - & 100 & 35 & 20 & 0 & 0 \\
20 & 20 & 20 & 20 & 0 & 0 & 75 & 100 & - & 20 & 20 & 0 & 0 \\
80 & 80 & 15 & 100 & 40 & 25 & 100 & 50 & 65 & - & 80 & 55 & 75 \\
30 & 30 & 20 & 45 & 0 & 85 & 85 & 85 & 85 & 30 & - & 0 & 20 \\
85 & 85 & 45 & 85 & 15 & 100 & 100 & 85 & 100 & 30 & 100 & - & 85 \\
65 & 40 & 40 & 40 & 0 & 0 & 100 & 65 & 80 & 60 & 80 & 40 & -
\end{bmatrix}$$

2. Determining the value in the discordance matrix required adding the weights included in the discordance set:

$$d_{kl} = \frac{\max\{|v_{kj} - v_{ij}|\} j \in D_{kl}}{\max\{|v_{kj} - v_{ij}|\} v_j}$$

So the calculation D_{12} was as follows:

$$D_{12} = \frac{\max\{|5,422-5,422|;|4,807-3,846|;|2,726-2-726|;|11,885-10,630|;|3,690-3,690|\}}{\max\{|5,422-5,422|;|4,807-3,846|;|2,726-2-726|;|11,885-10,630|;|3,690\ \ 3,690|\}}$$

$$= \frac{0}{0}$$

$$= 0$$

Calculation D_{13} arrived at D_{1312} done in the same way, so that the discordance matrix value was obtained:

$$D = \begin{bmatrix}
- & 0 & 1 & 0 & 1 & 0 & 0 & 0,366 & 0 & 0 & 0 & 1 & 0,146 \\
1 & - & 1 & 0 & 1 & 0 & 0 & 0,533 & 0 & 0 & 0 & 1 & 0,560 \\
0 & 0 & - & 0 & 1 & 0 & 0 & 0,154 & 0 & 0 & 0 & 0,385 & 0 \\
1 & 0 & 1 & - & 1 & 0 & 0 & 0,384 & 0 & 0 & 0 & 1 & 0,560 \\
0,916 & 0 & 0,697 & 0 & - & 0 & 0 & 0 & 0 & 0 & 0 & 0 & 0 \\
1 & 1 & 1 & 1 & 1 & - & 1 & 1 & 1 & 1 & 1 & 1 & 1 \\
1 & 1 & 1 & 1 & 1 & 0 & - & 1 & 0,575 & 1 & 1 & 1 & 1 \\
1 & 1 & 1 & 1 & 1 & 0,202 & 0,720 & - & 0 & 1 & 0,538 & 1 & 1 \\
1 & 1 & 1 & 1 & 1 & 0,217 & 1 & 1 & - & 1 & 1 & 1 & 1 \\
0,181 & 0,130 & 1 & 0,130 & 1 & 0 & 0 & 0,205 & 0,076 & - & 0,086 & 1 & 0,071 \\
1 & 1 & 1 & 1 & 1 & 0,086 & 0,490 & 1 & 0 & 1 & - & 1 & 1 \\
0,772 & 0,045 & 1 & 0,045 & 2,994 & 0 & 0 & 0,100 & 0 & 0,905 & 0 & - & 0,247 \\
1 & 1 & 1 & 1 & 1 & 0 & 0 & 0,698 & 0,324 & 1 & 0,418 & 1 & -
\end{bmatrix}$$

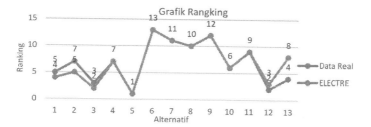

Figure 1. Ranking comparison.

3.4 *Ranking results*

From the dominant results of calculations using the ELECTRE method, the highest-ranking results were obtained, namely a fifth alternative named Anisa Hapsari, because it has a value of $E_{kl} = 1158.4$. The results of ranking the ELECTRE method were then compared with the results of the performance calculation undertaken in the DISPERINDAG. The process of comparison or validation was done using the Spearman Rank Correlation. The correlation calculation was based on the ranking generated by the ELECTRE method and the real data shown in Figure 1.

The Spearman Rank Correlation was 0.9368. Based on the correlation obtained, it can be said that the ELECTRE method produces output or information in accordance with real data. The closer to a value of 1, the better the correlation. This means that the model or method used approaches reality.

4   CONCLUSION

The conclusions obtained were the application of a decision support system using the ELEC-TRE method in a system for selecting the best employees. Survey results came from questionnaires based on a linked scale and based on the criteria of service orientation, integrity, commitment, discipline, and cooperation, where each criterion had sub-criteria. The results of the study obtained a Spearman Correlation Rank value of 0.9368. The closer to value 1, the better the correlation. This means that the model or method used approaches reality. So it can be said that the ELECTRE method can accurately rank employees' performance when using sub-criteria from the criteria of service orientation, integrity, commitment, discipline, and cooperation.

REFERENCES

Budinji, W. 2013. The measurement scale and the number of responses in the Likert scale. *Jurnal Ilmu Pertanian dan Perikanan*, 2(2), 125–131.

Cholil, S. R., & Indriyawati, H. 2019. Electre method for determining car stock at PT. New Ratna motor with a customer satisfaction approach. *Jurnal Transformatika*, 16(2), 160–168.

Fauzi, W. 2016. Sistem Pendukung Keputusan Penerima Bantuan Dana Rutilahu Dengan Menggunakan Metode Electre. *Seminar Nasional Teknologi Informasi dan Komunikasi*, 2016(Sentika), 2089–9815.

Fitriah, A., & Irfiani, E. 2018. Sistem Pendukung Keputusan Pemilihan Pegawai Terbaik PT Pegadaian Jakarta Dengan Metode Simple Additive Weighting. *Information System for Educators and Professionals*, 2(2), 105–114.

Helmi, T., Munjin, R. A., & Purnamasari, I. 2016. Kualitas Pelayanan Publik dalam Pembuatan Izin Trayek oleh Dllaj Kabupaten Bogor Effectiveness of Public Service in Service by Dllaj Route Permits Bogor District. *Jurnal Governansi*, 2(April), 47–59.

Maryuliana, Subroto, I. M. I., & Haviana, S. F. C. 2016. Sistem Informasi Angket Pengukuran Skala Kebutuhan Materi Pembelajaran Tambahan Sebagai Pendukung Pengambilan Keputusan Di Sekolah Menengah Atas Menggunakan Skala Likert. *TRANSISTOR Elektro dan Informatika*, 1(1), 1–12.

Nashrullah Irfan, M., Abdillah, G., & Renaldi, F. 2015. Sistem Pendukung Keputusan untuk Rekomendasi Promosi Jabatan ... (Nashrullah dkk.). *Sistem Pendukung Keputusan*, 5(1), 196–201.

Putra, A. A., Andreswari, D., & Susilo, B. 2015. Pinjaman Samisake Dengan Metode Electre. *Sistem Pendukung Keputusan untuk Penerima Bantuan Pinjaman Samisake dengan Metode Electre (Studi Kasus: LKM Kelurahan Lingkar Timur Kota Bengkulu)*, 3(spk), 1–11.

Engineering, Information and Agricultural Technology in the
Global Digital Revolution – Hendrawan & Wijayanti Dual Arifin (eds)
© 2020 Taylor & Francis Group, London, ISBN 978-0-367-33832-9

Study of the use of heat treatment in postharvest handling of red chili (*Capsicum annum* L.)

D. Larasati, E. Pratiwi & D.A. Gunantar
Universitas Semarang, Semarang, Indonesia

ABSTRACT: Red chili is a perishable commodity, so correct and appropriate postharvest cultivation is necessary. One of the postharvest handling technologies that can be used to prevent damage is heat treatment. The purpose of this study was to determine the effect of heat treatment on hardness, moisture content, vitamin C levels, and total red chili molds. The design of this study was a simple, utterly randomized design with hot water treatment (HWT). The treatment consisted of timed heat treatment for 0, 10, 20, 30, 40, and 50 minutes. The results of the study showed that the time of heat treatment had a significant effect on water content, vitamin C, texture, and total mold. The conclusion was that the length of time of heat treatment can reduce water levels and vitamin C but increase the texture of chili and inhibit the growth of mold.

Keywords: Heat Treatment, Red Chili

1 INTRODUCTION

Chili (*Capsicum annum* L.) is a perishable commodity, so in this case, postharvest handling acts as a chain unseparated from production activities. The success of the postharvest red pepper starts with the selection of seeds, planting, harvesting, postharvest, packaging, storage, and transportation, and continues with processing the results. Damage to red chili after harvest is usually caused by mechanical and physical issues and pests and diseases such as fruit flies (horses dense hand), anthracnose (Colletotrichum capsici sydow), and rot Phytophthora (Phytophthora capsici Leonian). Preventing damage to chili due to pests and diseases is necessary for excellent and correct postharvest handling. In postharvest handling, it is necessary to provide treatment that can maintain the chili's quality, one of which is heat treatment.

Heat treatment has become one of the pest and disease control technologies most widely performed on horticultural crops. In Gedong Gincu, mango fruit control has been carried out using the vapor thermal treatment (VTH) technique. Vapor thermal treatment with a temperature of 46.5°C followed by winding can maintain the mangoes for up to 28 days in storage (Marlisa, 2007). The results of the study showed that heat shock treatment is very influential in reducing chili injury (Pattiruhu, Purwanto, & Darmawanty, 2017) so that on this occasion, the researcher examined the effect of heat treatment on hardness, moisture content, vitamin C levels, and total mold on chili.

2 RESEARCH METHODS

2.1 *Materials and tools*

The material used in this study was money red chili taken from farmers in the Semarang district of central Java. We used some chemicals for analysis. The tools used were a water bath that provided heat treatment, incubators, autoclaves, and several glasswares for analysis.

2.2 Research methodology

The study was conducted to determine the changes in the quality of chili. We gave heat treatment by dipping chili fruit in a hot water bath with a temperature of 46–46.5°C with a length of time of 10, 20, 30, and 40 minutes, as well as controls (without heat treatment). We then observed the texture measured using a Rheometer model CR-300, water content (AOAC, 1984), vitamin C (Ranganna, 1997), and total mold (Suriawiria, 1990).

2.3 The design of the experiment

The study design was a simple, utterly randomized design with hot water treatment (HWT) with a medium temperature of 46°C, repeated four times. The treatment consisted of six levels. These were T0: without HWT, T1: HWT for 10 minutes, T2: HWT for 20 minutes, T3: HWT for 30 minutes, T4: HWT for 40 minutes, and T5: HWT for 50 minutes. The data were analyzed statistically by analysis of variance to see if differences occurred between treatments followed by a real-difference test using the Duncan Multiple Range Test (DMRT) at the level of 5%.

3 RESULTS AND DISCUSSION

3.1 Effect of timed heat treatment on chili water content

The red chili water content ranged from 12% to 15%; loss of water content caused loss of freshness, so the chili looked wilted, wrinkled, and unattractive (Winarno, Fardiaz, & Fardiaz, 1980). Variety analysis results showed that the time of heat treatment had a real, essential effect on chili water content. Chili water content from various treatments ranged from 11.30% to 13.52%.
Description:

1. The results comprised the average of five replications.
2. The mean followed by different superscripts showed significant differences in water content between treatments ($p < 0.05$).

Table 1 shows significant differences in water content between treatments. The longer the heat treatment, the lower the chili water content. This phenomenon is due to faster evaporation so that the product/chili loses water, which results in a decrease in the freshness of the chili that occurs in the fruit due to respiration rate increases, high air temperature, or humidity below 85–95% (Tranggono, 1990). The decrease in moisture content because of evaporation is due to the more prolonged heat treatment time; the heating time is getting longer, and respiration is getting faster.

3.2 Effect of timed heat treatment on chili vitamin C content

Vitamin C (ascorbic acid) in chili varies between 30 and 143.7 mg/100 g depending on the type or variety and the age of the chili. The chili used in this study was red chili TM, with a 100% maturity level. Vitamin C in fruit will decrease with faster transpiration.

Table 1. Chili water content.

Timed heat treatment	Water content (5)
T0 (without heat treatment)	13.13 ± 1.90 bc
T1 (10 minutes)	12.87 ± 1.56 bc
T2 (20 minutes)	13.52 ± 1.69 c
T3 (30 minutes)	12.30 ± 0.89 abc
T4 (40 minutes)	11.93 ± 2.88 a
T5 (50 minutes)	11.30 ± 1.48 a

Description:

1. The results comprised the average of five replications.
2. The mean followed by different superscripts showed significant differences in vitamin C content between treatments (p < 0.05).

Table 2 shows significant differences in vitamin C content between treatments. The longer heat treatments cause the vitamin C content to decrease because the heating time causes a faster speed of loss of moisture, while vitamin C is a water-soluble vitamin. According to Patria (2013), respiration is a chemical reaction of fruit that converts sugar, which is assisted by oxygen to carbon dioxide, water, and releases heat. Ferry's results (2018) show that blanching decreases vitamin C content. It is also said by Ferry (2018), that the decrease in vitamin C content in each water-blanching and steam blanching treatment is due to the warming effect when blanching which causes vitamin C is damaged degraded by heat so that vitamin C dissolves in water because it is easily oxidized.

3.3 *Effect of timed heat treatment on chili texture*

Changes in texture are one of the physiological changes that occur as a direct result of water loss in horticultural products. Fruit hardness can be measured qualitatively by taste or quantitatively using a rheometer.

Description:

1. The results comprised the average of five replications.
2. The mean followed by different superscripts showed significant differences in chili texture between treatments (p < 0.05).

Table 3 shows differences in the texture of chili produced between treatments. The difference in chili texture is influenced by the length of treatment time; the longer the heat treatment, the higher the value of the texture. This is due to the higher evaporation, the higher loss of water, which causes the product's flexibility to decrease. Heat treatment can also reduce the activity of enzymes that break down pectin on the cell walls of chili so that the fruit's hardness is higher than the treatment without supplying VTH at the same storage temperature (Vicente et al., 2005). The VTH treatment temperature of 47°C for 25–30 minutes did not cause damage to the fruit based on weight loss quality parameters, total dissolved solids, or

Table 2. Chili vitamin C content.

Timed heat treatment	Vitamin C (mg/100 g)
T0 (without heat treatment)	22.09 ± 1.14 c
T1 (10 minutes)	19.29 ± 1.31 b
T2 (20 minutes)	19.02 ± 1.52 b
T3 (30 minutes)	17.15 ± 1.83 ab
T4 (40 minutes)	16.05 ± 1.66 a
T (50 minutes)	16.15 ± 1.78 a

Table 3. Chili texture.

Timed heat treatment	Texture (g/mm)
T0 (without heat treatment)	111.90 ± 22.57 a
T1 (10 minutes)	114.79 ± 27.63 a
T2 (20 minutes)	124, 15 ± 25.96 a
T3 (30 minutes)	158.88 ± 28.13 b
T4 (40 minutes)	173.71 ± 20.06 b
T5 (50 minutes)	166.76 ± 14.95 b

Table 4. Total mold in chili (log cfu).

Timed heat treatment	Total mold (log cfu)
T0	6.76 ± 0.95 a
T1	5.57 ± 1.67 ab
T2	3.00 ± 2.04 ab
T3	3.45 ± 0.17 abc
T4	1.50 ± 1.73 c
T5	0.75 ± 1.50 c

hardness, and did not cause damage to physiology, where the pattern of climatic respiration was still reasonable (Lestari, Hasbullah, & Harahap, 2017).

3.4 *Effect of time heat treatment on total mold in chili*

Fungi is one of the microbes that cause postharvest damage to agricultural products in addition to bacteria. According to Ferry (2018), red chili that is still fresh is a fertile environment for microorganisms to grow and develop because of the high levels of moisture present on the surface. Microorganisms can grow faster in defective vegetables and fruits (Sopandi & Wardah, 2014).

Description:

1. The results comprised the mean of five replications.
2. The mean followed by different superscripts shows significant differences in mold in the chili between treatments ($p < 0.05$).

Table 4 indicates differences in the total chili molds from various heat treatments. The highest total mold was found in treatment T0 without heat treatment, and the lowest was found in treatment T5, which was 50 minutes, with a total mold of 0.75 ± 1.50 log cfu. It shows that the length of the heat treatment can control the total growth of molds. According to Lurie (2010), heat steam treatment with a temperature of 45°C can control fungi and insects, and a temperature of 60°C for 10 minutes can inhibit the growth of postharvest disease. The results of Spadoni et al. (2014) showed that a hot temperature of 60°C for 20 seconds effectively suppressed hyphae growth and conidia Monilia laxa in peaches.

Temperature and heating time affect mold growth. Fruit rot fungi generally grow optimally at temperatures of 20–25°C and minimum growth temperatures of 5–10°C (Utama, 2006). Ilmi, Poerwanto, and Sutrisno's (2015) research results, with washing temperatures of 53 ± 1°C and 60 ± 1°C, were not significantly different at 9 and 12 HSP but were significantly different from temperatures of 27 ± 1°C (average temperature).

4 CONCLUSION

We can conclude that Marlisas's (2007) timed heat treatment significantly affects water content, vitamin C, texture, and total mold in red chili (*Capsicum annum* L.). Pattiruhu et al.'s (2017) results showed that heat treatment resulted in water and vitamin C levels falling, texture increasing, and mold growth decreasing.

REFERENCES

AOAC. 1984. *Official Methods of the Association of Agricultural Analytical Chemists.* Arlington, VA: Association of Analytical Chemists, Inc.

Ferry. 2018. Effect of blanching on the quality of red chili (*Capsicum annum* L.). Thesis, Faculty of Agriculture, University of Muhammadiyah Yogyakarta.

Ilmi, N. K., Poerwanto, R., & Sutrisno. 2015. Treatment of hot water and temperature settings saved to maintain the quality of mangoes (*Mangifera indica* L.). *CV. Gedong J. Hort.*, *25*(1), 78–87.

Lestari, R., Hasbullah, R., & Harahap, I. S. 2017. Perlakuan Uap Panas dan Suhu Penyimpanan untuk Mempertahankan Mutu Buah Mangga Arumanis (*Mangifera indica* L.). *Jurnal Keteknikan Pertanian*, *5*(2).

Lurie, S. 2010. Postharvest heat treatments of horticultural corps. Department of Postharvest ARO Science, The Volcani Center Dagan Bet, Israel.

Marlisa, E. 2007. Study of disinfestation of fruit flies with vapour heat treatment in Gedong Gincu' manga. Thesis, Bogor Agricultural Institute, Bogor.

Patria, G. D. 2013. Perubahan Sifat Fisik dan Kimia Jambu Air (Syzygium Samarangense) Varietas Dalhari yang Dikemas Selama Penyimpanan Pada Suhu 5°C. Universitas Gadjah Mada.

Pattiruhu, G., Purwanto, Y. A., & Darmawanty, E. 2017. Perlakuan Panas untuk Mengurangi Gejala Kerusakan Dingin pada Mangga (*Mangifera indica* L.) var. Gadung selama Penyimpanan pada Suhu Rendah', *Comm. Horticulturae Journal*, *1*(1), 8–13.

Ranganna, S. 1997. *Handbook of Analysis and Quality Control for Fruits and Vegetable Products*. New Delhi: McGraw Hill.

Sopandi, T., & Wardah, A. 2014. *Mikrobiologi Pangan*. Yogyakarta: Penerbit Andi.

Spadoni, A., et al. 2014. Influence of hot water treatment on brown rot of peach and rapid fruit response to heat stress. *Postharvest Biology and Technology*, *94*, 66–73.

Suriawiria, U. 1990. *Pengantar Mikroba Umum*. Bandung: Penerbit Angkasa Bandung.

Tranggono, S. 1990 *Biokimia dan teknologi pasca panen*. Yogyakarta: PAU Pangan dan Gizi, UGM.

Utama, I. M. S. 2006. Control of post-harvest pest organisms for horticulture products in forming GAP: Empowering officers in the management of horticultural pests in order to support good agricultural practice (GAP).

Vicente, A. R., et al. 2005. Effect of heat treatments on cell wall degradation and softening in strawberry fruit. *Postharvest Biology and Technology*, *38*(3), 213–222.

Winarno, F. G., Fardiaz, S., & Fardiaz, D. 1980. *Introduction to Food Technology*. Gramedia.

Engineering, Information and Agricultural Technology in the
Global Digital Revolution – Hendrawan & Wijayanti Dual Arifin (eds)
© 2020 Taylor & Francis Group, London, ISBN 978-0-367-33832-9

Tuberculosis diagnosis using Bayes's theorem and a web-based forward chaining algorithm in Ungaran City

Henny Indriyawati, Prind Triajeng Pungkasanti & Wahyu Septiawan
Universitas Semarang, Semarang, Indonesia

ABSTRACT: Tuberculosis (TB) is an infectious disease caused by the *Mycobacterium tuberculosis* germ and easily transmitted through the air. Tuberculosis mostly attacks human organs such as the lungs, kidneys, intestines, and clear glands. Based on data from the RSUD Ungaran that examined patients at Ungaran Hospital, a lot of the disease that occurs in the city of Ungaran itself is TB. Another concerning piece of data reveals that the neonatal mortality rate in Semarang Regency in 2014 was 8.15 per 1,000, or as many as 74 cases. In response to these data, this study aimed to develop a TB diagnosis expert system for Ungaran using Bayes's theorem and a forward chaining algorithm. An expert system seeks to adapt human knowledge to computers so that computers can solve problems as human experts do. This expert system uses Bayes's theorem for calculating the probability value, and the forward chaining algorithm to produce a conclusion. This expert system is useful for diagnosing TB earlier, which can make it easier for sufferers to get proper treatment.

Keywords: Bayes's Theorem, Forward Chaining Algorithm, Expert System, Tuberculosis (TB)

1 INTRODUCTION

Ungaran City is located in the province of Central Java and borders the cities of Semarang and Demak Regencies in the north, Grobogan and Boyolali Regencies in the east, Magelang Regency in the south, and Tcmanggung and Kendal Regencies in the west (Decree of the Minister of Health of the Republic of Indonesia, 2014). The period shifting from the dry season to the rainy season or vice versa is called transition. Diseases that occur in tropical and subtropical areas are generally in the form of infections often called tropical diseases. Symptoms of emerging diseases are generally indicative of which disease is suffered. Based on data from the Semarang district of Dispendukcapil, at the end of 2014, the population of Semarang Regency was 989,399 people (Decree of the Minister of Health of the Republic of Indonesia, 2014). The number of newfound cases of tuberculosis (TB) in Ungaran in 2014 was 191 with a case notification rate (CNR) of 19.30 per 100,000 population, while the total number of cases up to 2014 was 286 with a CNR of 28.91 per 100,000 population. Based on data from the RSUD Ungaran examining patients at Ungaran Hospital, TB often occurs in the city of Ungaran itself; during the last six months of 2018, 70 cases of TB were found, of which pulmonary TB comprised about 0.8%, and extrapulmonary TB was found in glands (0–2%), bones (0.6%), and intestines (0.2%). In the developing world of information technology, expert systems can be utilized as one of the quick steps to identifying what diseases are attacking (Satzinger, Jackson, & Burd, 2011).

2 FOUNDATION THEORY

2.1 *Expert system*

"An expert system is a program that behaves like a human expert/expert (human expert)" (Siswanto, 2010). Expert systems are computer programs that solve complex real-world problems using computers that behave as human experts do when dealing with problems.

2.2 *Bayes's theorem*

Bayes's theorem was coined by Thomas Bayes around 1950. Bayes's theorem is a theory of probability conditions that calculates the probability of an event (hypotension) depending on another event (Rahayu, 2013).

2.3 *Tuberculosis (TB)*

Tuberculosis is an infection of the respiratory tract caused by the *Mycobacterium tuberculosis* bacterium (Rachmawati, 2015). This bacterium spreads through the saliva of TB sufferers who sneeze or cough.

2.4 *Symptoms of tuberculosis*

Symptoms of tuberculosis include:

1. Tuberculosis lung
2. Tuberculosis bone
3. Tuberculosis glands
4. Tuberculosis intestinal

2.5 *Analysis of the problem*

In current economic conditions, the cost of consulting a doctor about lung disease is not cheap; the small number of lung doctors also causes difficulty in obtaining a consultation regarding lung disease.

2.6 *Hardware requirements*

Hardware used in the making of this expert system included the Intel Core i3 2.4 GHz, 4 GB RAM, a 500 GB hard drive, a monitor, a keyboard, a mouse, and a printer.

2.7 *Software requirements*

Software components used to support the manufacturing of this expert system were as follows:

1. Operating system using Windows 10 Home Basic
2. Xampp to run the PHP programming language
3. MySQL as database

3 RESULTS AND DISCUSSION

1. IF you feel shortness of breath and chest pain (GJ01) AND you have a light fever sometimes with shivering at night (GJ02) AND you have coughed for more than three weeks

(GJ03) AND you feel weak (GJ04) AND you sweat at night (GJ05) AND your weight has dropped (GJ06) AND you have a blood-mixed phlegm cough/bloody cough (GJ07) THEN you might have TB in your lungs (HP01).

2. IF you have a light fever sometimes with shivering at night (GJ02) AND you have coughed for more than three weeks (GJ03) AND you are sweating at night (GJ05) AND you have lumps in your glands (neck, thighs, or armpits) (GJ08) AND the lumps feel painful when touched (GJ09) AND the lumps increase every day (GJ10) THEN you might have TB in your glands (HP02).

3. IF you have a light fever sometimes with shivering at night (GJ02) AND you sweat at night (GJ05) AND you have back pain (GJ11) AND you feel pain when bending (GJ12) AND you have a lump on your back or spine (GJ13) THEN you might have TB in your bones (HP03).

4. IF you have a light fever sometimes with shivering at night (GJ02) AND you have coughed for more than three weeks (GJ03) AND you sweat at night (GJ05) AND you have stomach pain (GJ14) AND you have difficulty defecating (GJ15) AND you often experience nausea (GJ16) AND you have an inflamed appendix (GJ17) THEN you might have TB in your intestines (HP04).

Analysis of Bayes's theorem

$$= \frac{(Geja1i\ selected * Ni1table\ probability)}{(Geja1total * Nilai\ Possible\ liquidity)} x100\%$$

Calculation example if some of the geology is used in one disease, with the geja code selected:

Nilai HP01: 0.80
   Gejala obtained by the user: GJ01, GJ04, GJ07
   Geotal disease, HP01: GJ01, GJ02, GJ03, GJ04, GJ05, GJ06, GJ07
   (GJ01 * HP01) + (GJ04 * HP01) + (GJ07 * HP01) / (GJ01 * HP01) + (GJ02 * HP01) + (GJ03 * HP01) + (GJ04 * HP01) + (GJ05 * HP01) + (GJ06 * HP01) + (GJ06 * HP01) + (GJ07 * HP01) =
   (0.80 * 0.80) + ((0.40 * 0.80) + (0.80 * 0.80)/(0.80 * 0.80) + (0.40 * 0.80) + (0.60 * 0.80) + (0.40 * 0.80) + (0.60 * 0.80) + (0.40 * 0.80) + (0.80 * 0.80)
   0.64 + 0.32 + 0.64
   = _____
   0.64 + 0.32 + 0.48+ 0.32 + 0.48 + 0.32 + 0.64
   1.6
   = _____
   3.2
   = 0.5 X 100% = 50%

The result is that the user experiences symptoms of TB of the lungs with nilai 50%. If the index approaches 100%, it indicates that the disease will occur.

4 RESULTS AND DISCUSSION

4.1 *Patient diagnosis page*

Figure 1 shows the results of the diagnosis of disease symptoms that have been previously selected and their initial treatment.

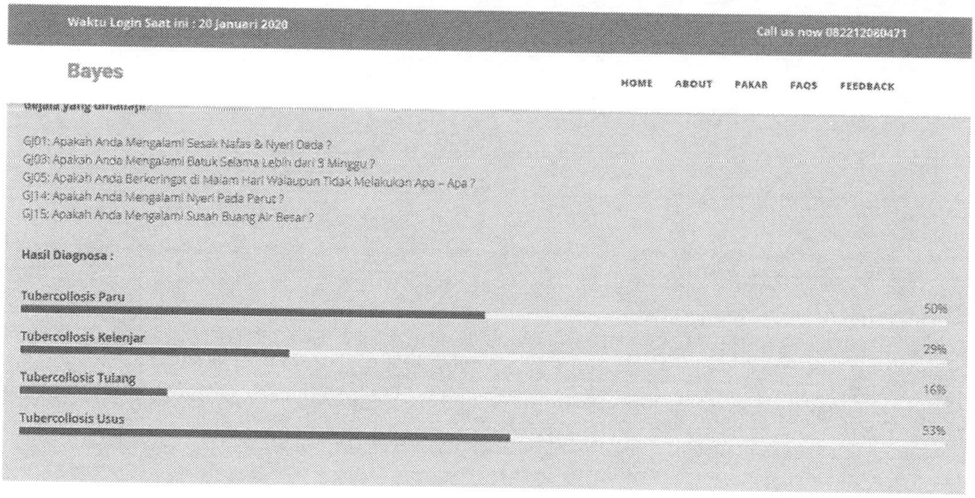

Figure 1. Patient diagnosis page.

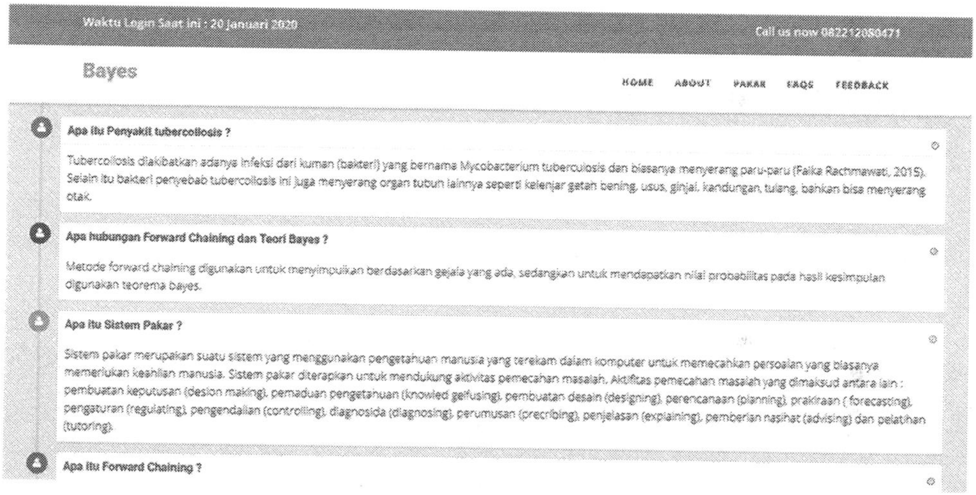

Figure 2. Patient FAQs page.

4.2 Patient FAQs page

Figure 2 shows an explanation of TB and of expert systems on a frequently asked questions (FAQ) page for patients.

4.3 Feedback page

Figure 3 shows a patient feedback page containing suggestions submitted by the user.

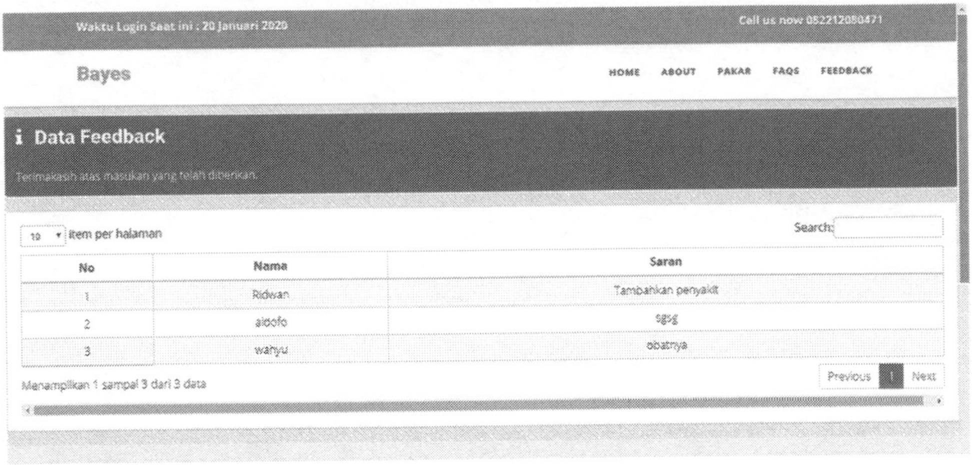

Figure 3. Patient feedback page.

5 CONCLUSION

Based on the results of the foregoing discussion, the author draws the following conclusions:

1. This system can provide accurate information and make it easier for people to know the early symptoms of TB in humans.
2. Bayes's theorem can be applied to diagnose TB in humans.
3. Based on the tests of this application carried out by pulmonary specialists, it can be concluded that this application can be used to diagnose patients because the system diagnosis results are 80% the same as doctors' diagnoses.

REFERENCES

Decree of the Minister of Health of the Republic of Indonesia. 2014. *Health Profile of Semarang Regency*.
Rachmawati, F. 2015. Prevalensi Penyakit Tuberculosis Paru di Kota Metro Provinsi Lampung Tahun 2011–2013. *Jurnal Biotek Medisiana Indonesia*, 4(1), 25–32.
Rahayu, S. 2013. Sistem pakar untuk mendiagnosa penyakit gagal ginjal dengan menggunakan metode bayes. *Pelita Informatika: Informasi dan Informatika*, 4(3).
Satzinger, J. W., Jackson, R. B., & Burd, S. D. 2011. *Systems Analysis and Design in a Changing World*. Boston, MA: Cengage Learning.
Siswanto. 2010. *Artificial Intelligence*. Yogyakarta: Graha Ilmu.

Engineering, Information and Agricultural Technology in the Global Digital Revolution – Hendrawan & Wijayanti Dual Arifin (eds)
© 2020 Taylor & Francis Group, London, ISBN 978-0-367-33832-9

Determining a motorcycle's PCE through a microsimulation based on driving behaviors

Iin Irawati & Supoyo
Universitas Semarang, Semarang, Indonesia

ABSTRACT: Passenger Car Equivalent (PCE) is a greatly important quantity used to determine traffic flow, and it is utilized to convert the number of vehicles per hour into Passenger Car Units (PCU) per hour. Indonesia has heterogeneous traffic characteristics with various types of vehicles and undisciplined driving behaviors reflected in high aggressiveness levels on the roads. Passenger car equivalent quantities must reflect existing traffic conditions. The PCE of the *Indonesian Highway Capacity Manual* (IHCM) therefore must be updated since the 1997 IHCM was considered a product of traffic data processed in 1994. This study aimed to create a traffic simulation model to determine a motorcycle's PCE based on driving behaviors using VISSIM. The research location was along the Majapahit signalized intersections, Semarang, Central Java. The research variables consisted of the number of vehicles per hour, geometric intersection, time cycle, and driving behaviors. The outcome indicator was queue length. Several standardized values of driving behavior were examined based on field conditions, such as calibration, including average standstill distance = 0.5 m; additional safety distance = 0.8 m; multiplied safety distance = 1 m; minimum headway = 0.5 m; and safety distance reduction factor = 0.6. The result of the motorcycle's PCE output was 0.15.

Keywords: Signalized Intersection, Driving Behavior, Heterogeneous, PCE, Traffic Flow

1 INTRODUCTION

The *Indonesian Highway Capacity Manual* (IHCM) (1997) is a guideline to analyze signalized roads and intersections in Indonesia. The field conditions indicate that the result of road segment analysis using IHCM 1997 were greater than that in actual field conditions. However, the results of analysis on signalized intersections are smaller than those in field conditions. One factor influencing performance is traffic flow. Traffic flow plays an important role in determining current conditions and estimating future traffic conditions (Kafy et al., 2018). Traffic flow is formed by multiplying the volume and the passenger car equivalent (PCE) value. The performance estimation may become more accurate if the traffic volume is considered as one fundamental measure in traffic analysis so that its mixture is converted into traffic flow by multiplying the value of PCE (Zahiri & Chen, 2018). Thus, PCE has a critical part in determining traffic flow. Scholars have used several variables in previous research to determine the PCE value, including: headway (Jili et al., 2012; Mamun et al., 2012), current flow and traffic density (Gani, Yoshii, & Kurauchi, 2017), queue discharge flow (Praveen & Arasan, 2013), speed (Zerjawi & Razzaq, 2016), saturation flow (Skabardonis et al., 2014), delay (Nassiri, Tabatabaie, & Sahebi, 2017), V/C (Fan, 1990), vehicle-hour (Metkari et al., 2012), travel time (Keller & Saklas, 1984; Miller, 1968), rainfall intensity (Alhassan & Ben-Edigbe, 2012), saturation flow and loss time (Van Zuyler & Li, 2010), and weather (Zahiri & Chen, 2018).

Based on these studies, the researchers tried to determine the value of the PCE gap variable from the previous research, that is, the driving behavior variable. Conditions in the field indicated that drivers' behaviors influence traffic flow, which then influences the PCE value. Lane maneuvers that are part of the driving behaviors in different variations of traffic density have a significant impact on traffic flow characteristics (Zhu et al., 2018). The PCE values were designed in developed countries with homogeneous traffic signals, so they are not appropriate in heterogeneous traffic conditions. Indonesia has various types of vehicles and undisciplined driving behaviors, as shown by a high aggressiveness level reflected in drivers' tendency to enter gaps in the middle of red signalized intersections. Characteristic differences between homogeneous and heterogeneous traffic flow make it necessary to conduct a study on the performance formation value of intersections in heterogeneous countries – especially Indonesia – using an approach based on drivers' behaviors. A microsimulation method was chosen to determine the PEC value while the device used to build the model was a VISSIM. The research was conducted at Majapahit signalized intersections in Semarang, Central Java, Indonesia.

2 RESEARCH METHODOLOGY

Methodology involves stages through which to conduct a study. The research methodology flowchart can be seen in Figure 1.

3 ANALYSIS

The data analysis stages were as follows:

1. Build a microsimulation model with VISSIM. The stages were summarized as follows: (1) Make background images based on locations in Google Map and link connectors by

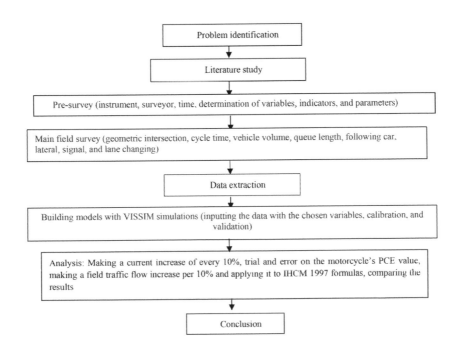

Figure 1. Research methodology flowchart.

entering the geometric intersections. (2) Enter the vehicle type based on the vehicle class in the field. (3) Make a travel route for the static vehicle routing by first entering the vehicle volume in the vehicle input (number of vehicles per hour: west = 3,306, east = 3,078, south = 1,518, north = 1,540. (4) Make the vehicle composition (for example from the west: MC = 0.8, LV = 0.17, HV = 0.02, UMC = 0.01). (5) Make a signalized head and then set the signalized traffic counting the signalized control based on the number of cycle times (for example, in the first phase by first selecting the west direction, then the red time = 121 seconds, green = 45 seconds, yellow = 3 seconds). (6) Make a data and queue counter checklist to evaluate the configuration. (7) The model calibration is based on the driving behavior characteristics for the heterogeneous traffic in the menu of Wiedemann 74. (8) Run the program with VISSIM simulation, for which the results are shown in Figure 2.

2. Conduct trial and error on the motorcycle's PCE value from 0.19 to 0.15 by inputting it into the IHCM 1997 formula. The basis of the proposal is to improve motorcycles' PCE in Indonesia, as motorcycles represent the highest number and greatest proportion of vehicles in Indonesia.

3. Make an increase in the field of traffic flow per 10% and apply it to the IHCM 1997 formula with the motorcycle's PCE value of 0.15. The results are shown in Figure 3.

4 DISCUSSION

Several measurable components in the field can be used as benchmarks for drivers' behaviors in an intersection: following cars (stalking vehicles), lane changes (switching paths), and lateral movement or pull. The parameters of drivers' behavior at intersections, including metropolitan cities such Semarang in Indonesia, are classified into average standstill distance = 0.5 m; additional safety distance = 0.8 m; multiplied safety distance = 1 m; minimum headway = 0.5 m; and safety distance reduction factor = 0.6 with a random distribution pattern (any).

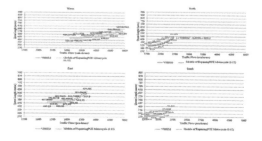

Figure 2. Calibration models (following car) and VISSIM simulation results (queue length and 3D models).

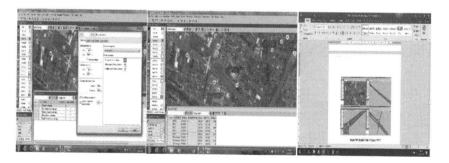

Figure 3. Comparative results of models in the fields with a motorcycle PCE value of 0.15.

Based on the quantitative measure, drivers' behaviors are used to determine PCE. Motorcycles' PCE was one proposed improvement for IHCM 1997 since their number dominated heterogeneous traffic in Indonesia. The PCE value greatly influences traffic flow. The higher the PCE value, the greater the current that may be produced. The nonproportional capacity value will affect the degree of saturation (DS) value. The higher the DS, the higher the queue that may be produced. In IHCM 1997, motorcycles' PCE value was 0.2. The value of 0.2 was chosen based on data taken in 1994, which are significantly different from current conditions. To adjust for current conditions, the proposed IHCM 1997 repair is 0.15.

5 CONCLUSION

The results of signalized intersection performance with the implementation of a PCE value of 0.2 indicate a significant difference between field queue length and IHCM 1997. Using the motorcycles' PCE value of 0.15 implies that if motorcycle volume is multiplied by the PCE value of 0.15, it will give a smaller current when compared to the PCE value of 0.2. The smaller the current produced, the more the queue length may decrease. The PCE value of 0.15 was chosen as the proposed improvement of IHCM 1997 because it produced the queue length closest to the field queue length.

REFERENCES

Alhassan, H. M., & Ben-Edigbe, J. 2012. Evaluation of passenger car equivalent values under rainfall. In *2012 International Conference on Traffic and Transportation Engineering (ICTTE 2012)* (pp. 6–10). Singapore: IACSIT Press.

Fan, H. S. L. 1990. Passenger car equivalents for vehicles on Singapore expressways. *Transportation Research Part A: General*, 24(5), 391–396.

Gani, F. A., Yoshii, T., & Kurauchi, S. 2017. Estimation of the dynamic value of passenger car units in mixed traffic. *Journal of the Eastern Asia Society for Transportation Studies*, 12, 1665–1675.

Indonesian Highway Capacity Manual. 1997.

Jili, X. et al.2012. Study on the passenger car equivalent at a signalized intersection. In *2012 Fifth International Conference on Intelligent Computation Technology and Automation* (pp. 490–493).

Kafy, A.-A. et al.2018. Estimating traffic volume to identify the level of service in major intersections of Rajshahi, Bangladesh. *Trends in Civil Engineering and Its Architecture*, 2(4), 292–309.

Keller, E. L., & Saklas, J. G. 1984. Passenger car equivalents from network simulation. *Journal of Transportation Engineering*, 110(4), 397–411.

Mamun, A. A. et al.2012. Estimation of passenger car equivalent (PCE) of right-turning vehicles at signalized intersections in Dhaka Metropolitan City, Bangladesh. *Advances in Transportation Studies*, 26.

Nassiri, H., Tabatabaie, S., & Sahebi, S 2017. Delay-based passenger car equivalent at signalized intersections in Iran. *Promet-Traffic & Transportation*, 29(2), 135–142.

Praveen, P. S., & Arasan, V. T. 2013. Influence of traffic mix on PCU value of vehicles under heterogeneous traffic conditions. *International Journal for Traffic & Transport Engineering*, 3(3).

Skabardonis, A. et al.2014. Developing improved truck passenger car equivalent values at signalized intersections. *Transportation Research Record*, 2461(1), 121–128.

Van Zuyler, H. J., & Li, J. 2010. Measuring the saturation flow, lost time, and passenger equivalent values of signalized intersection. *Traffic and Transportation Studies*, 312–329.

Zahiri, M., & Chen, X. 2018. Measuring the passenger car equivalent of small cars and SUVs on rainy and sunny days. *Transportation Research Record*, 2672(31), 110–119.

Zerjawi, A., & Razzaq, A. K. 2016. Estimation of free flow speeds and passenger car equivalent factors for multilane highways. *International Journal of Scientific & Engineering Research*, 6(7).

*Engineering, Information and Agricultural Technology in the
Global Digital Revolution – Hendrawan & Wijayanti Dual Arifin (eds)
© 2020 Taylor & Francis Group, London, ISBN 978-0-367-33832-9*

Solar power inverter in Photo-voltaics with full bridge systems

Harmini, Andi Kurniawan Nugraha & Titik Nurhayati
Universitas Semarang, Semarang, Indonesia

ABSTRACT: Solar power inverter is one of the core components in the PLTS system in order to produce the power or energy needed by the load. Solar power inverter is utilized to convert from DC energy (Direct Current) to AC (Alternating Current) to load. In this research, a simulation of a solar power inverter system on a photovoltaic system will be designed. Simulation needs to be done using SIMULINK-MATLAB software. Solar power inverter is developed using a boost converter as a controller and a Full Bridge Inverter. Boost converter is designed from 12VDC - 17 VDC to 300 VDC. Full-Bridge Inverter produces AC voltage of 240 VAC with modulation index 0.8 and frequency 60 Hz. Total Harmonic Distortion is 46.07%. Power of Photovoltaic Solar panels generated at 3500 watts by 1000 watt/m2 irradiation conditions and temperature temperatures of 25 degrees Celsius.

Keywords: Solar Power, Inverter, Boost, Photovoltaic, Full Bridge

1 INTRODUCTION

Electrical energy generated by a Photo-voltaic system requires a control system, and an inverter called a solar power inverter (Kshirsagar and Vadirajacharya, 2014). Inverters can generate electricity that is required by household electrical loads (Czarkowski, 2001). The control system is required so that the energy produced can adjust the energy required for the load. The electrical energy generated by the sun always changes according to the conditions of radiation and the temperature of the sun (Ho and Chung, 2005). A solar power inverter is one of the main components of the solar power system to produce the power or energy needed by the load. Solar power converts from DC (Direct Current) electricity to AC (Alternating Current) (Reinoso *et al.*, 2012). This makes solar power inverters and control systems very important in the PLTS system (Seel, Barbose and Wiser, 2013).

In this study, a simulation of a solar power inverter system on a photovoltaic system will be designed. Simulation is done using SIMULINK-MATLAB software. The solar power inverter that will be designed has pure-sine feature and has proper output regulation according to the input of the inverter. A solar power inverter is developed using a boost converter as a controller and a full-bridge inverter. The simulation results will be applied to the actual hardware. The problem formulation in this study is how to design and simulate a solar power inverter on a Photovoltaic system. The purpose of this study is to create and affect solar power inverters on photovoltaic systems, which are looking forward to being used as a reference for making real hardware.

2 RESULT AND ANALYSIS

2.1 *Solar power inverter circuit*

The design of a Solar Power Inverter consisting of Boost Converter and the design of a Full-Bridge Inverter (Prabaharan *et al.*, 2017). The voltage released by Photovoltaic Solar Panels is

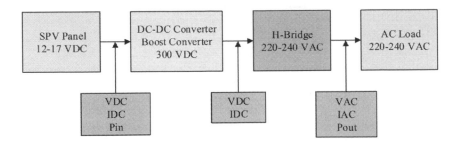

Figure 1. Block diagram of Solar Power Inverter.

12-17 Volt DC, this voltage will be raised by DC-DC Converter with Boost Converter topology of 300 Volt DC, then the voltage will be converted to AC voltage of 220 frequencies
50Hz according to the load requirements used for household appliances. In general, the design
diagram block is shown in Figure 1.

2.2 *Boost converter circuit*

The design of the boost converter in this study is the 12-17 Volt DC input voltage output from
solar photovoltaic panels will be increased to 300 Volt DC voltages and 8A current as the
input voltage and current in the Inverter. The switching frequency used in PWM (Pulse Width
Modulation) is 1080 Hz. The design and output of the boost converter circuit are shown in
Figure 2 and Figure 3.

a. Determine the value of the duty cycle

$$\frac{V_o}{V_i} = \frac{1}{1-D}$$

$$\frac{300}{12} = \frac{1}{1-D}$$

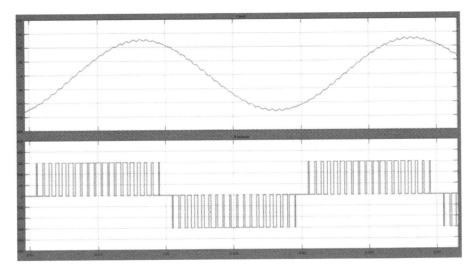

Figure 2. Boost Converter Simulation.

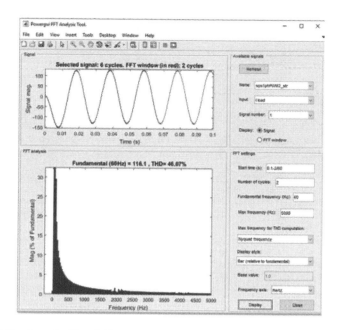

Figure 3.   Full Bridge Inverter Simulation.

$$D = 1 - \frac{12}{300}$$

$$D = 0.96$$

b. Determine R value nilai R

$$Load\ Resistance\ (R) = \frac{Vo}{Io}$$

$$Load\ Resistance\ (R) = \frac{300\,V}{8A}$$

$$Load\ Resistance\ (R) = 37,5\ Ohm$$

c. Determine Inductor value

$$L = \frac{(1 - D)^2 D\ x\ R}{2f}$$

$$L = \frac{(1 - 0,96)^2 0,96\ x\ 37,5}{2\ x\ 1080}$$

$$L = 26\,mikro\,Henry$$

The inductor value installed in the circuit is 1.25 times greater than the calculation value, so the value of the inductor is 1.25 x 26 microHenry which is 32.5 microHenry.

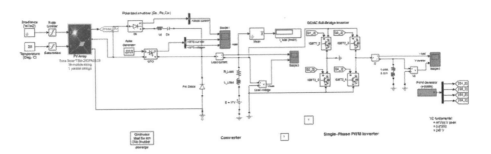

Figure 4.   Load Current Wave and Full Bridge Inverter Voltage.

d. Determine capasitor value

$$C_{\min} = \frac{DV}{\Delta v_0 Rf}$$

$$C_{min} = \frac{0,96 \ x \ 300}{0,001 \ x \ 37,5 \ x \ 1080}$$

$$C_{min} = 0,071 \, Farad$$

Based on Figure 2, it can be seen that the load current generated by the boost converter circuit can produce a load current of 160 Ampere and an output voltage of 300 Volt DC. This Boost Converter output voltage will be the Full-Bridge Inverter input voltage. The simulation of the Full-Bridge Inverter design is shown in Figure 5; the simulation is done using IGBT. Each IGBT is triggered by a PWM signal with a 1080 Hz switching frequency. The modulation index is 0.8, and the AC frequency is 60 Hz.

The fundamental voltage produced by the inverter is the magnitude of the modulation index multiplied by the peak voltage input from the inverter, which is 0.8 times the 300 VDC equal to 240 VAC. The inverter simulation results are shown in Figure 5, which is in the form of current and inverter full bridge voltage. In the picture, the load current has formed a sine wave, but the sine signal produced still contains ripple or noise. The load used is not only resistive but also inductive. Meanwhile, the output voltage is always a square wave pulse or square wave. This happens because the inverter still produces an output voltage with a lot of frequency, so a filter is needed to show the output voltage with the desired frequency.

The inverter output voltage analysis is seen from the review of THD (Total Harmonic Distortion) shown in Figure 5. From this picture, there is a distortion or loss of inverter output wave with a fundamental signal wave of 46.07%. This means that the ripple that occurs is very large, so a filter is needed to suppress the ripple value or damage the inverter output wave signal.

The total power that can be generated reaches 3498.04 Watts or 3500 Watts in 1000 Watts/m2 irradiation conditions and STC temperature (Standard Temperature Condition) of 25 degrees Celsius, with the number of SPV panels 14 units with a series of 1 parallel and 14 series. The overall series of Solar Power Inverter simulations is shown in Figure 6.

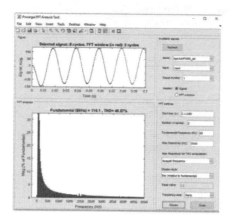

Figure 5.   THD Analysis (Total Harmonic Distortion).

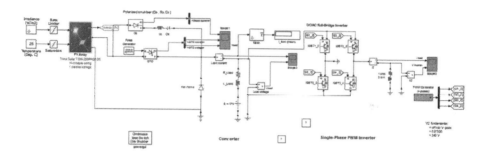

Figure 6.   The overall series of Solar Power Inverters.

3   CONCLUSION

The Solar Power Inverter system consists of a Boost Converter system and a Full Bridge Inverter. Boost Converter is designed from 12VDC - 17 VDC to 300 VDC. Full Bridge Inverter produces AC voltage of 240 VAC with modulation index 0.8 and frequency 60 Hz, Total Harmonic Distortion of 46.07%. Photovoltaic Solar panel power generated by 3500 Watts in 1000 Watt/m2 irradiation conditions and temperature of 25 degrees Celsius.

REFERENCES

Czarkowski, D. (2001) *DC-DC Converter in Power Electronic Handbook*. University of Florida.
Ho, B. M. T. and Chung, H.H. (2005) 'An Integrated Inverter with Maximum Power Point Tracking For Grid-Connected PV system', *IEEE Trans Power Electron*, pp. 953–962.
Kshirsagar, J. and Vadirajacharya, K. (2014) 'One-Cycle-Controlled Single-Phase Inverter for Grid Connected PV System', *Advance in Electronic and Electric Engineering*, 4(5), pp. 455–462.
Prabaharan, N. *et al.* (2017) 'Integration of single phase reduced switch multilevel inverter topology for grid connected photovoltaic system', *Energy Procedia*. Elsevier, 138, pp. 1177–1183.
Reinoso, C. R. S. *et al.* (2012) 'Photovoltaic inverters optimisation', *Energy Procedia*. Elsevier, 14, pp. 1484–1489.
Seel, J., Barbose, G. and Wiser, R. (2013) 'Why are residential PV prices in Germany so much lower than in the United States', *A scoping analysis, presentation, Lawrence Berkeley National Laboratory. download under* http://eetd. lbl. gov/node/50173.

Engineering, Information and Agricultural Technology in the
Global Digital Revolution – Hendrawan & Wijayanti Dual Arifin (eds)
© 2020 Taylor & Francis Group, London, ISBN 978-0-367-33832-9

Smart door lock and automatic lighting device with bluetooth connection using android Arduino

April Firman Daru, Susanto, Aria Hendrawan & Atmoko Nugroho
Universitas Semarang, Semarang, Indonesia

ABSTRACT: In the current era of technology, humans are increasingly busy, and many tools have been created that can facilitate human life. An obstacle has arisen, however, in regulating electronic devices used in the home, such as devices involved in the automation of doors and home lighting. Given this background, researchers have designed a tool for remote door automation and home lighting using Bluetooth so that it can be controlled using an Android application. This system combines hardware and software such as the Arduino UNO R3 as the main controller, and is supported by other modules such as the HC-05 module, relay, DF mini-player, speaker, and servo motors.

1 INTRODUCTION

Science and technology, especially in the field of information technology, have developed very quickly in recent years. Along with the technological development of the times, communication problems are increasingly becoming very complex. Communication devices such as the Android smartphone can be used not only for periodic communication but also for utilizing newly developed technological features. Android is an operating system for cell phones based on Linux. Android provides an open platform for developers to create their applications for use by various mobile devices, and it is commonly used on smartphones and tablet personal computers (PCs). It functions the same as the Symbian operating system on Nokia, iOS on Apple, and BlackBerry OS (Safaat, 2015). One of the advantages of an Android smartphone is that it is easy to do programming on it, and it can be connected to a microcontroller such as Arduino. A microcontroller is a computer device that contains a chip used to control electronic equipment per preprogrammed instructions. It can be called a "small controller" where an electronic system that previously needed many supporting components such as TTL and CMOS ICs can be reduced and finally centralized and controlled by this microcontroller (Syahwil, 2013). Arduino is a microcontroller family name board created by the Smart Projects company, one of the creators of which is Massimo Banzi. Arduino is an "open-source" hardware that can be made by anyone. Arduino programming is done through computers using a software called Arduino Integrated Development Environment (Arduino IDE) (Abdul Kadir, 2013). Integrated Development Environment uses Java for writing programs, compiling binary code, and uploading the code into the microcontroller's memory. The Arduino IDE window consists of three main parts. The top part, the toolbar, comprises a menu file, edit and sketch tools, and a help function. The middle part contains the program code. The bottom part is a message window or consul that provides status and error messages (Wicaksono, 2017).

The design of this system was tested on the Arduino UNO R3, with this tool expected to facilitate daily activities, especially in terms of managing door locks and home lighting. Arduino UNO is an ATmega328-based microcontroller board with 14 digital input–output (I/O) pins (where 6 pins can be used as output PWM), 6 analog inputs, a clock speed of 16 MHz, a Universal Serial Bus (USB) connection, a power jack, a header ICSP, and a reset button. The board uses power connected to a computer with a USB cable or external power with an AC-DC adapter or battery (Syahwil, 2013). Arduino UNO has specifications such as the Arduino UNO board, an ATmega328 microcontroller, 5V operating voltage, 7V–12V input voltage, 6V–20V input voltage limit, 14 I/O digital pins (6 output pins PWM), 6 analog input pins, 40mA DC current per I/O pin, DC current for 3.3V 50mA pin, flash 32 KB memory (ATmega328), where 0.5 KB is used by a bootloader, 2 KB SRAM (ATmega328), 1 KB EEPROM (ATmega328), and a 16 MHz clock.

Research around Bluetooth and Arduino as controllers has also been published in a journal article entitled "Bluetooth System Implementation Using Android and Arduino for Electronic Equipment Control." But the research featured in the article was limited to television with a distance of 320 cm and a width of 180.4 cm (Rahmiati, Firdaus, & Fathorrahman, 2014). In this study, the author discusses the controllers' use in locking and unlocking doors.

System design focuses on constructing a reliable system. The method used in writing this research was the prototype method. The prototype begins with identifying the needs of the software. Then a prototype program is made so that it can tested against what is desired from the system (Pressman, 2012).

The device under examination in this research requires a human actor who controls a tool using Android to turn on or turn off lights and open or close doors.

2 RESULT AND DISCUSSION

2.1 *Diagram block*

A diagram block is very important in the design of a system, because it provides a simple description of the whole system.

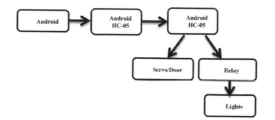

Figure 1.   Diagram block.

The smart door lock and lighting system under study here works as follows:

1. Android gives commands to turn on/off lights and to open or close doors.
2. The HC-05 module functions to connect Android and Arduino.
3. Arduino UNO serves as a platform for entering program commands and as the brain of the system.
4. The servo opens and closes doors.
5. The relay connects and disconnects an electrical current in order to control lights.

2.2 *Flowchart design system*

A flowchart is a graphical presentation that describes a step-by-step solution to a problem. Flowcharts are used not only to describe simple operations also to deal with complex problems.

Compared to a pseudocode, flowcharts have the advantage of being a reliable method of communication, using only a few symbols that are easily understood by anyone. The shape reflects the real situation; for example, a flowchart can describe repetition or branching, and errors can be detected visually (e.g., steps that have not been directed to another step) (Abdul Kadir, 2013).

Following is an explanation of Figure 2:

1. Start.
2. Log in to the application.
3. Check whether the password is correct.
4. Connect your HP/Android with Bluetooth.
5. If successful, the Android and Arduino devices will connect and a text notification will appear.
6. Give a voice command to turn on the light.
7. If the voice command is correct according to the keyword, the Arduino process will be carried out and forwarded by the relay so the light will turn on. If the command is wrong, the user will be told to repeat the voice command with the correct keyword.
8. Give a voice command to turn off the light.
9. If the voice command is correct according to the keyword, it will be processed by Arduino and forwarded by the relay so that the electrical power to the light will be disconnected and the light will turn off. If the command is wrong, the user will again be told to repeat the voice command with the correct keyword.
10. Give a voice command to open the door.
11. If the voice command is correct according to the keyword, the process will be carried out by Arduino and forwarded by the servo so the door will automatically open. If the command is wrong, the user will again be told to repeat the voice command with the correct keyword.
12. Give a voice command to close the door.
13. If the voice command is correct according to the keyword, it will be processed by Arduino and forwarded by the servo so that the door will be closed and locked automatically. If the command is wrong, the user will again be told to repeat the voice command with the correct keyword.
14. End.

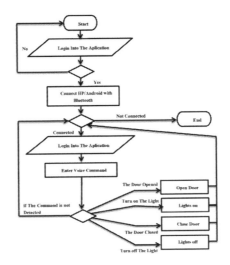

Figure 2.   Flowchart.

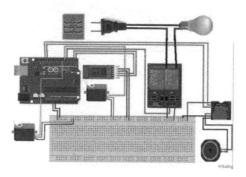

Figure 3.   Making a prototype.

2.3   *Making a prototype*

The following schematic depicts the entire prototype that the author has designed. The prototype consists of an Arduino board module, a Bluetooth module, a relay, a DF player module, a servo motor, and interconnected loudspeakers. Following is an explanation of Figure 3:

1. Arduino UNO is a program for processing the microcontroller that has been created to run existing sensors.
2. The HC-05 Bluetooth module uses bridge communication to send and receive data between smartphones and Arduino.
3. Relay modules function as electrical circuit-breakers and current connectors. The relay is used for controlling lights.
4. Electric current from the socket powers the lamp.
5. The DF player mini-module is used as an MP3 player for bookmarks or activity notifications on the device while running.
6. The loudspeaker performs as the media output sound from the DF player.

2.4   *Software implementation*

After the circuit is complete and the tool can run correctly, then an Android application is made using the MIT Inventor app. This application sends commands that are integrated with commands on the microcontroller via Bluetooth. This application comprises a number of buttons, including a button to connect to Bluetooth, a button to display the maker information, and a microphone button to input voice commands. The Bluetooth button serves to select and connect Bluetooth from Android to Arduino. The microphone button in the layout functions as a medium for inputting voice commands.

3   CONCLUSION

The conclusion of this research is that the smart door lock and home lighting system using Android-based Arduino can be a solution to home lighting and security problems often faced by residents. Home security features can be added to the system in order to make it more modern. But this prototype can only be operated with a distance between the phone and the Bluetooth module of less than 60 meters.

4 PHOTOGRAPHS AND FIGURES

Suggestions from this research include adding security features such as alarms and closed-circuit television (CCTV) cameras that are equipped with motion sensors. The communication bridge between the smartphone and Arduino, which originally used Bluetooth, can be updated using a Wi-Fi module.

REFERENCES

Abdul Kadir, A. 2013. *Practical Guide to Studying Microcontroller Applications and Its Programming Using Arduino.* Yogyakarta: Andi.

Pressman, R. S. 2012. *Software Engineering: Practitioner Approach.* 7th Edition. Yogyakarta: Andi.

Rahmiati, P., Firdaus, G., & Fathorrahman, N. 2014. Implementasi Sistem Bluetooth menggunakan Android dan Arduino untuk Kendali Peralatan Elektronik. *ELKOMIKA: Jurnal Teknik Energi Elektrik, Teknik Telekomunikasi, & Teknik Elektronika,* 2(1), 1.

Safaat, N. H. 2015. *Pemograman aplikasi mobile smartphone dan tablet PC Bahasa Android.* Bandung: Informatika Bandung.

Syahwil, M. 2013. *Panduan Mudah Simulasi dan Praktek Mikrokontroler Arduino.* Yogyakarta: Penerbit Andi.

Wicaksono, M. F. H. 2017. *Mudah Belajar Mikrokontroler Arduino.*

*Engineering, Information and Agricultural Technology in the
Global Digital Revolution – Hendrawan & Wijayanti Dual Arifin (eds)
© 2020 Taylor & Francis Group, London, ISBN 978-0-367-33832-9*

Using asphalt recycling (RAP) for road passion layers

Supoyo & Agus Muldiyanto
Universitas Semarang, Semarang, Indonesia

Faizal Mahmud
*Civil Engineering Study Program, Faculty of Engineering, Sultan Agung Islamic University, Semarang,
Indonesia*

ABSTRACT: The advancement of technology in constructing highways is increasing with the
advent of faster, more efficient road-building techniques. One such current technology is the use
of reclaimed asphalt pavement (RAP). This is a used hot-mix material resulting from the exca-
vation of cold milling mixed with dolomite additives to improve the quality and longevity of the
asphalt. The technology is considered effective and also reduces damages that often occur after
repairs. This research, therefore, aimed to determine the Marshall, VIM, VMA, and VFB
values of each mixing variation. In this study, the percentages of dolomite used as mixture were
2%, 4%, 6%, 8%, and 10% of the asphalt weight. Based on the results, it was conclusively found
that the mixture showed an increase in the VIM and VFB values while increasing the dolomite
percentage. However, the VMA value remained unchanged, hence it was concluded that the
addition of dolomite affected asphalt filling between the material cavities.

Keywords: Reclaimed Asphalt Pavement (RAP), Dolomite, VIM Value, VMA Value, VFB
Value

1 INTRODUCTION

The importance of good roads to daily human activities cannot be overemphasized even
though roads are often damaged and require repair or reconstruction. This has necessitated
a periodic maintenance plan using overlay (Lubis & Mochtar, 2008). Overlay is a current technology used for making highways and is often employed in the process of making or repairing
asphalt roads. This technique is achieved by using a mixture of certain additives to improve
the quality of the asphalt.

Flexible pavement consists of lower and upper foundations and surface layers. The latter is
in the form of a mixture of asphalt with coarse and fine aggregate (Departemen Pekerjaan
Umum, 2007). The unification process is carried out at a certain temperature followed by the
comparison of the asphalt, coarse, and fine aggregate, which has been determined through
mix design. The strength and durability of the road pavement itself are largely determined by
the weight-carrying capacity of the soil and the type of asphalt used, as well as the aggregates
that are the main material (Rahman, 2010). The purposes of this study were to review and
analyze the reuse of reclaimed asphalt pavement (RAP) in pavement layers, determine the percentage of dolomite that can be added in a paved mixture, ascertain the mixed asphalt levels
with aggregate RAP, and determine the characteristics of the asphalt mixture from the aggregate results of the RAP.

Based on the mixing method, the RAP is divided into recycling: cold recycling – for
example, cement-treated recycling base (CTRB), cement-treated recycling base-base
(CTRBB), mixture with emulsion asphalt binder, mixture with liquid asphalt binder, Bitumen

Foam – and recycling hot mixes – for example, recycling heated scrap material AMP or mixing in place.

The fatigue performance of wear-coated asphalt concrete mixtures using recycled materials and styrene-butadiene-styrene polymers was examined by Novita, Subagio, and Rahman (2011). The test results revealed that RAP has lower penetration and higher viscosity. The testing showed that increasing the proportion of RAP raises the fatigue resistance of the mixture. Mochtar (2012) examined the optimization of the use of cold milling materials for base course coating mixtures, via the cement-treated recycled base method. Gradation of the extracted RAP then indicated a mismatch of the desired specifications (Bina Marga V), and this incompatibility can be corrected with an aggregate redundant blending (Bina Marga, 1998). However, the quality of the asphalt contained in RAP still meets the requirements of asphalt penetration 60/70. Kasan (2009) examined the recycled concrete mixtures by adding rejuvenating ingredients. From the results of testing the extraction of old asphalt mixture materials, the asphalt content in the mixture was determined to be 4.6% such that the estimated variations were 5.6%, 6.1%, and 6.6% with a variation of 0% fluxing material; 5%, 10%, 15%, and 20%, respectively. The results then showed that the more rejuvenating ingredients in the mixture caused a decrease in the mixed density and stability. The maximum flux material content that still meets the Marshall stability and other remaining specifications was 35.29% and 46.18%, respectively.

Suroyo (2004) examined the effect of recycling asphalt chunks on the properties of concrete. The results showed that mixtures using old chunks of aggregate were better than Job Mix AC values, and that it could also be reused. Meanwhile, Susilowati (2000) studied the use of used oil residues (ROB) as a rejuvenating material for concrete asphalt in recycling pavement. The results showed that optimum bitumen content of 9.634% and asphalt concrete mixture meet the requirements of Bina Marga (1987), and could be used for medium-class traffic. This study therefore aimed to introduce one method of road pavement repair that utilizes recycled old pavement materials, to obtain the physical and mechanical properties of hot asphalt concrete mixtures as a result of recycling from the old pavement, and to determine optimum content in the recycled hot asphalt concrete mixtures from the old pavement.

2 RESEARCH METHODS

2.1 Scope of research

This research included preparation and testing of raw materials and was the preliminary research on extraction, aggregate sieve analysis, and old pavement type weight testing. A mixture of hot asphalt and dolomite with variations of 0%, 2%, 4%, 6%, 8%, and 10% of the asphalt weight were prepared. Furthermore, stability and melt tests were carried out.

Table 1. Requirements for aggregate foundations.

	Class A	Class B
Rough aggregate abrasion (AASHTO T96-74)	0–40%	0–40%
Plasticity index (AASHTO T90-70)	0–6	4–10
Liquid limits (AASHTO T 98-68)	0–25	0–35
The product of the plasticity index with the percentage of passing the 75 micron filter	25 max.	
The soft part (AASHTO T 112-78)	0–5%	0–5%
CBR (AASHTO T 193)	90 min.	60 min.

Source: Direktorat Jenderal Bina Marga (2010)

2.2 Research variables

The independent variable in this study ranged from 5% to 7%, varying by 0.5% for asphalt grade, and 0.25%, 0.5%, 0%, 75%, and 1% of chunks of asphalt concrete. The dependent variable or research parameters consisted of stability and melt. This included density and percent cavity in aggregate, percent cavity in the mixture, cavity percentage filled with asphalt, melt, stability, and Marshall quotient. The materials used in this study were asphalt pavement or waste.

Table 2. Requirements for gradation of the aggregate layer material.

Sieve Size (mm)	Percent Weight Pass		Sieve Size (mm)	Percent Weight Pass	
	Class A	Class B		Class A	Class B
50	100	100	4.75	29–44	25–55
37.5	100	88–95	2	17–30	15–40
25	79–85	70–85	0.425	7–17	8–20
9.5	44–58	30–65	0.075	2–8	2–8

3 RESULTS AND DISCUSSION

Included in this study was the aggregate gradation, and the percentage of asphalt mixture was at 4%. The results of the plain asphalt level (Pb) based on the formula mix calculation design were at 5.83%. This value was, however, greater than the RAP extraction results such that the addition of asphalt was needed as a binder. In this study, the aggregate was not added while only asphalt with variations of dolomite mixture of 0%, 2%, 4%, 6%, 8%, and 10% was added to the asphalt mixture.

3.1 Air cavity analysis in mixtures (VIM)

The more asphalt content increased in the mixture, the more the VIM value decreased, and this means that the cavity in the mixture was small. Not enough space was available to accommodate a possible rise in asphalt to the surface. The amount of asphalt that can fill the cavity between grains became larger; hence, the volume of cavities in the mixture decreases while making asphalt concrete more durable.

Table 3. Results of density analysis.

% (VIM)	0%	2%	4%	6%	8%	10%	Rata-rata
	4.05	5.26	21.32	11.11	5.35	28.15	12.54

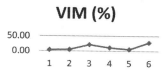

Figure 1. Graph of the relationship between asphalt levels and VIM.

3.2 Inter-aggregate air cavity analysis (VMA)

The more asphalt content increased in the mixture, the lower the VMA value became, because the amount of asphalt entering the cavity was not enough to fill the cavity. But the experiments showed that the VMA values were constant when added to the dolomite mixture. This then indicated that the asphalt content was not affected by the increase.

Table 4. Results of analysis of air spaces between aggregates.

% (VMA)	0%	2%	4%	6%	8%	10%	Rata-rata
	62.25	62.25	62.25	62.25	62.25	62.25	62.25

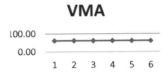

Figure 2. Graph of the relationship between asphalt levels and VMA.

3.3 Analysis of asphalt-filled air cavity (VFB)

This analysis showed that VMA values decrease when asphalt levels increase. This is due to the increasing amount of asphalt that can fill the cavity in the mixture. In addition to the increasing VFB value, it also shows more asphalt-covered aggregates.

Table 5. Analysis of asphalt-filled air cavity.

% (VFB)	0%	2%	4%	6%	8%	10%	Rata-rata
	93.50	91.55	96.93	96.53	96.31	97.20	95.33

Figure 3. Graph of the relationship between asphalt levels and VFB.

4 CONCLUSION

The analysis showed that mixing RAP and asphalt with dolomite additive material increases the VIM and VFB values in the aggregate cavity. Hence when added with gradual additives, dolomite was obtained as a percentage of the filling of the tight cavity. This greatly affects the durability of the asphalt and aggregate mixture.

REFERENCES

Bina Marga. 1998. Spesifikasi Umum Proyek Rehabilitasi/Pemeliharaan Jalan dan Jembatan Propinsi DIY. Departemen Pekerjaan Umum. 2007. Spesifikasi Umum Bidang Jalan dan Jembatan. *Pusat Litbang Prasarana Transportasi Badan Penelitian dan Pengembangan.*

Direktorat Jenderal Bina Marga. 2010. Spesifikasi Umum Bina Marga Divisi 6, Perkerasan Aspal.

Kasan, M. 2009. Studi Karakteristik Volumetrik Campuran Beton Aspal Daur Ulang. *SMARTek*, 7(3).

Lubis, Z., & Mochtar, B. 2008. Evaluasi Rumusan Damage Factor (equivalent axle load) dalam Perancangan Sistem Perkerasan Lentur Jalan Raya Akibat Adanya Muatan Berlebihan. *Jurnal Teknologi dan Rekayasa Sipil Torsi. Surabaya.*

Mochtar, I. B. 2012. Optimalisasi Penggunaan Material Hasil Cold Milling Untuk Campuran Lapisan Base Course Dengan Metode Cement Treated Recycled Base. *Jurnal Teknik Pomits*, 1(1), 1–6.

Novita, P., Subagio, B. S., & Rahman, H. 2011. Kinerja Kelelahan Campuran Beton Aspal. *Jurnal Transportasi*, 11(3), 163–172.

Rahman, H. 2010. Evaluasi Model Modulus Bitumen Asbuton dan Model Modulus Campuran yang Mengandung Bitumen Asbuton. Program Doktor Teknik Sipil, Institut Teknologi Bandung.

Suroyo, H. 2004. Pengaruh Daur Ulang Bahan Bongkahan Aspal terhadap Sifat-sifat Beton Aspal (Studi Kasus di Jalan Gajahmada Tegal). Tesis Program Pasca Sarjana Universitas Diponegoro.

Susilowati. 2000. Pemanfaatan Residu Oli Bekas sebagai Bahan Peremajaan untuk Daur Ulang Perkerasan Jalan. *Makalah Seminar Jurusan Teknik Sipil Politeknik Negeri Jakarta.*

Engineering, Information and Agricultural Technology in the
Global Digital Revolution – Hendrawan & Wijayanti Dual Arifin (eds)
© 2020 Taylor & Francis Group, London, ISBN 978-0-367-33832-9

Bandwidth management decision support system with hybrid (SAW and AHP) method

W. Adhiwibowo, B.A. Pramono, S. Hadi & N. Hidayati
Universitas Semarang, Semarang, Indonesia

ABSTRACT: The development of internet technology requires very large bandwidth. Semarang University, which has the second largest number of PTS students in Central Java, of course requires a great capacity for bandwidth as well. The Ministry of Research, Technology and Higher Education's standards state that students need bandwidth above 0.75 Kbps. This bandwidth is also for the use of lectures conducted at Semarang University. With limited bandwidth quotas, a decision support system was created to manage the allocations of bandwidth in each department at Semarang University, prioritized according to need. This decision support system was built using a hybrid method, namely the Analytic Hierarchy Process (AHP) and the Simple Additive Weighting (SAW) approach, where AHP was used for the weighting process. The aim of using this method was to reduce decision makers' subjectivity, while SAW was used for ranking processes; the result was the final preference value. The best alternative was chosen based on the highest preference value. The results obtained from this study included the determination of the priority scale for faculties that have bandwidth requirements so there is no mistake in determining the allocation of bandwidth.

Keywords: SPK, Hybrid, AHP, SAW, Bandwidth

1 INTRODUCTION

The use of technology to support lecture activities is very basic. With the second largest number of PTS students in Central Java in 2018, Semarang University is committed to improving the quality of facilities and infrastructure in the field of technology for lectures. One focus of these needs is obtaining the bandwidth required for internet access. These needs must be fulfilled because the minimum rule of good bandwidth requirements is 0.75 Kbps, according to the Ministry of Research, Technology and Higher Education (DIKTI). This need does not include online (network) recovery.

Changes in the lecture method applied by DIKTI to become a lecture model online (Minister of Education and Culture of the Republic of Indonesia, 2013) require sufficient technological resources. Data communication and appropriate topology models are needed for online activities. The inability of resources networks will hinder online activities held by universities. The aim of DIKTI online, often called SPADA (Indonesian Online Learning System), is to increase equity in learning carried out by universities in Indonesia. This is done because universities still lack human resources.

To manage online learning, it is necessary to manage the network so that online activities will be conducted by Semarang University.

Based on these requirements, the researchers were interested in designing a decision support system to determine bandwidth allocation according to the specified criteria, in order to facilitate the implementation of university policies in an appropriate, fast, effective, and efficient manner.

Before the allocation of bandwidth, static allocation was used to share bandwidth between networks. Of course, static allocation of bandwidth has advantages and disadvantages. Based on the research on static bandwidth allocation, several weaknesses emerged in the static allocation of bandwidth (Lashkari, Zeidanloo, & Sabeeh, 2011), namely lack of service support, weakness of certain data traffic, problems with quality of service, and bandwidth demand inflation.

This research combined Simple Additive Weighting (SAW) and Analytical Hierarchy Process (AHP). This method was chosen because it could discern the best alternative.

2 LITERATURE REVIEW

Research on bandwidth allocation in networks with many services has produced guaranteed and fair bandwidth allocation for its users and for newly installed services (Fei, Tian, & Lian, 2012). An optimal bandwidth allocation study of point-to-multipoint wireless networks has resulted in a balanced network load during quiet and busy times (Wu et al., 2015). Another study has examined an allocation of bandwidth that depends on usage. The study sought a balanced and dynamic approach to producing the maximum bandwidth.

Heterogeneous network research wants mobile network users to be connected continuously without any deterioration in network quality (Liang & Yu, 2018). The simulation generated from the research resulted in the best network selection based on the services used.

Scholars conducting recent research on bandwidth allocation have utilized a hybrid fuzzy AHP algorithm to determine bandwidth allocation on heterogeneous wireless networks (Goyal, Kaushal, & Sangaiah, 2018). The latest research regarding optimal network selection has employed algorithms such as SAW and MEW (Singh, 2015).

3 METHODOLOGY

The completion of the AHP and SAW method algorithms was as follows:

1. Step 1: Define the criteria that will serve as benchmarks for solving problems and determine the importance of each criterion.
2. Step 2: Calculate the comparison matrix value of each criterion based on the table of importance.
3. Step 3: Calculate the criteria weight value (Wj).
4. Step 4: Calculate the preference weight value (Vi).
5. Step 5: Rank these values.

Seven faculties at Semarang University were analyzed in order to examine the allocation of internet bandwidth quota.

Table 1. Faculty property for each alternative (data 2018–2019).

Alternative	FACULTY						
	FTIK	PSYCHOLOGY	LAW	TECHNIC	ECONOMY	THP	MASTER
Number of Comp. Lab.	11	0	0	3	4	0	1
Number of Lecturers	62	22	26	63	130	19	17
Number of Students	3,108	1,099	1,231	3,083	6,386	799	562
Weekly Teaching Time	75	60	60	60	75	60	75
Number of Study Programs	3	1	1	3	3	1	2

4   RESULTS AND DISCUSSION

Step 1

Determine the priority scale of each criterion. In this case, based on the evaluation results of the research team, K3 (number of computer labs) and K4 (number of lecturers) were the top priority, followed by K1 (number of students) and K5 (teaching time in one week) as the second priority, and K2 (number of study programs) as the last priority. These problems can be decomposed into a priority ladder.

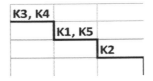

Figure 1.   Priority ladder.

Calculate the pairwise matrix value (paired comparison matrix) of each criterion. The following is the pairwise comparison matrix table of the aforementioned criteria.

Table 2.   The first iteration.

	Number of Comp. Lab.	Number of Lecturers and Employees	Number of Students	Weekly Teaching Time	Number of Study Programs
Number of Comp. Lab.	6.0000	5.0000	14.6667	13.6667	31.0000
Number of Lecturers and Employees	6.9333	5.9333	18.0000	17.0000	37.0000
Number of Students	2.2667	1.9333	5.3333	5.0000	11.6667
Weekly Teaching Time	2.2667	1.9333	5.3333	5.0000	11.6667
Number of Study Programs	1.0222	0.8222	2.4000	2.2000	4.6000

Table 3.   Both iterations.

	Number of Comp. Lab.	Number of Lecturers and Employees	Number of Students	Weekly Teaching Time	Number of Study Program
Number of Comp. Lab.	166.5778	139.9333	403.5111	376.8667	844.1556
Number of Lecturers and Employees	199.8933	167.9600	483.9556	452.0222	1013.0000
Number of Students	62.3526	52.3748	151.1556	141.1778	316.0222
Weekly Teaching Time	62.3526	52.3748	151.1556	141.1778	316.0222
Number of Study Programs	26.9630	22.6652	65.3659	61.0681	136.9378

Table 4.   Results and average.

Results	Average	Eigen 2
1.6899	5.2569	0.3215
2.0275	5.2570	0.3857
0.6328	5.2568	0.1204
0.6328	5.2568	0.1204
0.2738	5.2544	0.0521
		1.0000

Results = first matrix (1) multiplied by the second Eigen value iteration
Average = yield/Eigen 2

Table 5.   Results, phi, CI and CR.

	Results
15.2568	
	phi
1.0514	
	CI
−0.987160518	
	CR
0.8.8139332	

Step 4
Calculate the preference weight with the SAW method.

$$\mathbf{V1} = (1.6899 * 11) + (20,275 * 62) + (0.6328 * 3,108) + (0.6328 * 75) + (0.2738 * 3) = \mathbf{2159.250471}$$

$$\mathbf{V2} = (1.6899 * 0) + (20,275 * 22) + (0.6328 * 1,099) + (0.6328 * 60) + (0.2738 * 1) = \mathbf{778.2694513}$$

$$\mathbf{V3} = (1.6899 * 0) + (20,275 * 26) + (0.6328 * 1,231) + (0.6328 * 60) + (0.2738 * 1) = \mathbf{869.906313}$$

$$\mathbf{V4} = (1.6899 * 3) + (20,275 * 63) + (0.6328 * 3,083) + (0.6328 * 60) + (0.2738 * 3) = \mathbf{2122.447585}$$

$$\mathbf{V5} = (1.6899 * 4) + (20,275 * 130) + (0.6328 * 6,386) + (0.6328 * 75) + (0.2738 * 3) = \mathbf{4359.540316}$$

$$\mathbf{V6} = (1.6899 * 0) + (20,275 * 19) + (0.6328 * 799) + (0.6328 * 60) + (0.27\,38 * 1) = \mathbf{582.3534}$$

$$\mathbf{V7} = (1.6899 * 1) + (20,275 * 17) + (0.6328 * 562) + (0.6328 * 75) + (0.2738 * 2) + (0.2738 * 2) = \mathbf{439.78528}$$

Step 5

Table 6. Ranking of hybrid method.

No	Alternative name	Value weighting preferences	Information
1	FTIK	2,159.250471	rank 2
2	PSYCHOLOGY	778.2694513	rank 5
3	LAWS	869.906313	rank 4
4	ENGINEERING	2,122.447585	rank 3
5	ECONOMIC	4,359.540316	rank 1
6	THP	582.3534019	rank 6
7	POST	439.7852797	rank 7

5 CONCLUSION

This study concluded, through the application of a hybrid AHP and SAW method, that the number one ranking for the most bandwidth quota allocation was the Faculty of Economics, followed by FTIK, then the Faculty of Engineering, the Faculty of Law, the Faculty of Psychology, the THP Faculty, and finally postgraduates. Further research can apply bandwidth management allocation under the results of this study. Subsequent development determines the amount of bandwidth allocation for each faculty at Semarang University.

REFERENCES

Fei, W., Tian, H., & Lian, R. 2012. Utility-based dynamic multi-service bandwidth allocation in heterogeneous wireless networks. In *2012 IEEE 75th Vehicular Technology Conference (VTC Spring)* (pp. 1–5).

Goyal, R. K., Kaushal, S., & Sangaiah, A. K. 2018. The utility based non-linear fuzzy AHP optimization model for network selection in heterogeneous wireless networks. *Applied Soft Computing, 67,* 800–811.

Lashkari, A. H., Zeidanloo, H. R., & Sabeeh, A. A. 2011. Static bandwidth allocation on optical networks. In *International Conference on Machine Learning and Computing* (pp. 498–503).

Liang, G., & Yu, H. 2018. Network selection algorithm for heterogeneous wireless networks based on service characteristics and user preferences. *EURASIP Journal on Wireless Communications and Networking, 2018*(1), 241.

Minister of Education and Culture of the Republic of Indonesia. 2013. Permendikbud No. 109 of 2013 Implementation of Distance Education in Higher Education.

Singh, N. P. 2015. Optimal network selection using MADM algorithms. In *2015 2nd International Conference on Recent Advances in Engineering & Computational Sciences (RAECS)* (pp. 1–6).

Wu, M. et al. 2015. Evaluation of novel intelligent wireless pushing mechanism based on AHP. In *2015 Seventh International Conference on Advanced Computational Intelligence (ICACI)* (pp. 79–83).

Engineering, Information and Agricultural Technology in the Global Digital Revolution – Hendrawan & Wijayanti Dual Arifin (eds)
© 2020 Taylor & Francis Group, London, ISBN 978-0-367-33832-9

Fuzzy EAS optimization for Soybean production analysis

Khoirudin, Nur Wakhidah & Astrid Novitaputri
Universitas Semarang, Semarang, Indonesia

ABSTRACT: Determining the capacity for production in agriculture is helpful for farmers' jobs. Production may be predicted based on harvest areas and people's consumption needs. With a soft computing method, we may obtain the right production optimization by employing the fuzzy EAS method, a combination of fuzzy logic and genetics used to improve performance based on the mutations generated from the fuzzy evolution (Fuzzy EAS) algorithm. The number of production is 187,992 while the number of consumption is 2,012,712. In this case, these numbers were utilized to determine the optimal number of production solutions. The data used in this study showed that production from 2010 to 2015 experienced many imbalances in terms of soybean stock production, and 83% of genetic mutations were discovered from this study. The scope of this research was Central Java, Indonesia, with the aim of optimizing soybean production and consumption for the next generation.

Keywords: Production, Consumption, Agriculture, Fuzzy, Genetics

1 INTRODUCTION

Agriculture is an activity utilizing biological resources that humans process in order to produce food, industrial material, or energy sources to manage the environment (Hobbs, 1996). One type of food currently farmed in Asia is the soybean. The results of the National Social and Economic Survey (Susenas), conducted by the Indonesian Central Bureau of Statistics (BPS) in 2015, showed that the average consumption of soybeans per person per year was 7.51 kg while the fulfillment of soybean needs was only around 67.28%. With high soybean consumption and lack of supplies, the government has had to rely on imports from the United States, Brazil, and Argentina. In addition to minimum soybean production, another arising issue is the insufficiency of product availability in several regions. This study focused on Central Java Province (BPS, 2017; Data Pertanian, 2015).

In the field of information technology, these problems are considered challenges by the Agricultural Office, and a holistic approach should be taken to production that employs supply chain management, an umbrella between products and consumers that maximizes value. This study focused on soybean production and consumption (Prihatman, 2000).

2 METHODOLOGY

This research examined soybean per capita consumption and production using data retrieved from www.data.go.id (Data Pertanian, 2015). Figure 1 explains the fuzzy EAS algorithm flows (Nhita, 2016; Nhita, Wisesty, & Ummah, 2015).

Figure 1.   Fuzzy EAS algorithm flow chart.

3   RESULTS AND DISCUSSION

3.1   *A troubleshooting cycle using genetic and fuzzy algorithm*

3.1.1   *Determining the population initialization*

The chromosomal initialization used in producing the initial population in this system consisted of three factors: production and consumption, harvest area, and genetic initialization for chromosomal formation. The selected research data were only from Central Java and comprised 18 genes of six chromosomes (2010–2015) (Pertanian, 2015).

Table 1.   Formation of the supporting genes (solution details).

Year	Soybean Production	Harvest Area	Soybean Type Consumption
2010	187,992	114,070	2,897,973
2011	112,273	81,988	3,060,173
2012	152,416	97,112	2,930,413
2013	99,318	65,278	2,941,227
2014	125,467	72,235	4,363,180
2015	129,794	70,629	4,509,160

Table 2.   Population examples.

Line No.	
Individual 1	[2010 187,992 114,070 2,897,973; 2011 112,273 81,988 3,060,173; 2012 152,416 97, 112 2,930,413; 2013 99,318 65,278 2,941,227; 2014 125,467 72,235 4,363,180; 2015 129,794 70,629 4,509,160]
Individual 2	[2010 112,273 114,070 3,060,173; 2011 187,992 97,112 2,897,973; 2012 152,416 81, 988 2,941,227; 2013 99,318 70,629 2,930,413; 2014 125,467 72,235 4,509,160; 2015 129,794 65,278 4,363,180]
. . . .	. . . .
Individual n	[2010 187,992 81,988 2,941,227; 2011 112,273 114,070 4,363,180; 2012 152,416 97,112 4,509,160; 2013 99,318 72,235 2,897,973; 2014 125,467 70,629 3,060,173; 2015 129,794 65,278 2,930,413]

3.1.2 Evaluate fitness evaluation

In this request, the fitness values may determine the number of constraint violations to optimize. The constraints used to optimize this production included:

a. No consumption less than production, penalty is 1.
b. No production less than consumption, penalty is 1.

$$f(g) = \frac{1}{(1 + \sum Pi)}$$

P is the number of penalties given to each rule i.
The following is an example of population fitness calculation:

3.1.3 Selection

During selection, the fitness value was assessed using a rank selection. As a result, the best chromosome quality has a greater possibility of being selected in the next generation/iteration. The selected chromosomes will be used as the carriers to crossover.

3.1.4 Fuzzy (inference)

This inference process was divided into two parts: inference for probability of recombination (Pc), and inference for probability of mutation (Pm). The method used in this process was an inference model, where the resulted outputs were in the form of fuzzy sets containing linguistic values and membership degrees.

The Mamdani method used for the fuzzy EAS algorithm consisted of two inputs and two outputs. The two inputs were the population and the number of generations. The two outputs were the probability of recombination value (Pc) and the probability of mutation value (Pm) (Fachrie, Widowati, & Hanuranto, 2012).

Table 3. Universe of talk, domain population, and generation domains.

Population		Generation	
Universe of talk	[0 1,000]	Universe of talk	[0 1,000]
SMALL domain	[50 250]	SMALL domain	[50 200]
MEDIUM domain	[80 275]	MEDIUM domain	[80 275]
LARGE domain	[350 500]	LARGE domain	[350 500]

The probability of recombination (Pc) values were generally between 0.6 and 0.9 in the universe of talk, and the domains for the output results were the probability of recombination values. The used value rules were as follows (Fachrie et al., 2012):

Table 4. Probability of recombination and probability of mutation.

Probability of Recombination		Probability of Mutation	
Universe of talk	[0.6 0.9]	Universe of talk	[0 0.25]
SMALL domain	[0.625 0.7]	SMALL domain	[0.025 0.1]
MEDIUM domain	[0.63 0.7 0.72 0.78]	MEDIUM domain	[0.047 0.083 0.1 0.14]
LARGE domain	[0.72 0.78 0.8 0.87]	LARGE domain	[0.1 0.14 0.167 0.2]
Universe of talk	[0.8 0.875]	Universe of talk	[0.15 0.225]

Using two data consisting of population 18 and generation 6, the probability of recombination value was 0.70 and the probability of mutation value was 0.217.

99

3.1.5 *Crossover genetics*

The chromosomal selection study used for crossover was randomly conducted. The chromosomes coming to the crossover iteration were those with probability values greater than those of the specified crossover with 0.70. Chromosomes with greater probability values than those of the crossover will form new individuals.

3.1.6 *Mutations*

In this study, mutations were conducted by randomly selecting the chromosomes. Mutations were used to prevent information loss from exchanging information in the form of genes in the chromosomes.

3.2 *Implementation of fuzzy eas*

Repetition conditions are completed if after finding the production data optimization, we calculate performance, optimize selection based on the consumption attributes, generalize the liners and use fuzzy conditions obtained from Pc and Pm values, and then weight them to data resulting in the value of 2,012,172, as shown in Figure 2.

3.3 *Production period calendar*

The data obtained from www.data.go.id in 2010–2015 (Pertanian, 2015), processed using fuzzy EAS algorithms, determined the feasibility. If consumption was too high, then it was said to be improper. Based on the mutations obtained from the fuzzy EAS, production was 187,992 while consumption was 2,012,712.

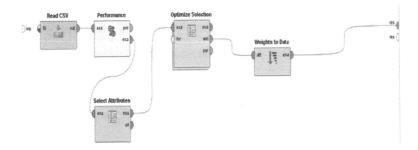

Figure 2. Implementation of fuzzy EAS 1.

Table 5. Data collection matrix.

Type	True	False
High consumption	12	2
Low consumption	1	3

$$\text{Recall} = \frac{12}{12 + 2} = 0.86$$

$$\text{Precision} = \frac{12}{12 + 3} = 0.8$$

$$F - \text{Measure} = \frac{Recall * Precision}{Recall + Precision} = \frac{2 * 0.86 * 0.8}{0.86 + 0.8} = \frac{1,376}{1.66} = 0.83$$

These results showed that the respective values of recall accuracy were 86%, precision was 80%, and F-Measure of production was 83%. This means that the system has a quite high accuracy in identifying whether consumption is high or low based on the large existing number of high consumption data (Kim, 2003).

4 CONCLUSION AND SUGGESTION

Based on the existing data examining production from 2010 to 2015, many imbalances occurred in soybean production. The results also showed that the genetic mutations have reached 83% since the mutations were generated from the Fuzzy EAS. The best mutation was production number 187,992 and consumption number 2,012,712. Soybean production did not meet soybean consumption in Central Java, so the next-generation factors and harvest areas need to be developed. Further research may be developed related to these applications.

REFERENCES

Data Pertanian. 2015. *Outlook Komoditas Pertanian Tanaman Pangan Kedelai. Kementrian Pertanian*, www.data.go.id.

Fachrie, M., Widowati, S., & Hanuranto, A. T. 2012. Implementasi Fuzzy Evolutionary Algorithms Untuk Penentuan Posisi Base Transceiver Station (BTS). *Jurnal Fakultas Hukum UII*. Islamic University of Indonesia.

Hobbs, J. E. 1996. A transaction cost approach to supply chain management. *Supply Chain Management: An International Journal, 1*(2), 15–27.

Kim, K. 2003. Financial time series forecasting using support vector machines. *Neurocomputing, 55*(1–2), 307–319.

Nhita, F. 2016. Implementasi Algoritma Weighted Moving Average Pada (Fuzzy Eas) Untuk Peramalan Kalender Masa Tanam Berbasis Curah Hujan. *Indonesian Journal on Computing, 1*(1), 77–94.

Nhita, F., Wisesty, U. N., & Ummah, I. 2015. Planting calendar forecasting system using evolving neural network. *Far East Journal of Electronics and Communications, 14*(2), 81.

Prihatman, K. 2000. *Tentang Budidaya Pertanian: Kedelai*. Deputi Menegristek Bidang Pendayagunaan dan Pemasyarakatan Ilmu Pengetahuan dan Teknologi.

Engineering, Information and Agricultural Technology in the Global Digital Revolution – Hendrawan & Wijayanti Dual Arifin (eds)
© 2020 Taylor & Francis Group, London, ISBN 978-0-367-33832-9

Goal frame detection system using image processing with LabVIEW

I. Udlhiya, B. Supriyo, E.D. Wardihani & R.M. Firdaus
Politeknik Negeri Semarang, Semarang, Indonesia

ABSTRACT: The advancement of digital image processing technology can facilitate human life, and this technology has many potential applications for various fields, such as robotics technology. The aim of this study was to design and develop a detection system for the goal frame in the soccer robot field based on digital image processing techniques. An image processing technique transforms the input image into an output image so that the output image has better quality. By utilizing the appropriate technology, a specific object can be identified based on the object capture from the camera. In the field of soccer robots, the robot should have the ability to identify the goal frame accurately. In this case, the robot used the camera as an image sensor, myRIO as a data acquisition system board, and LabVIEW as a programming tool. The results showed that the proposed technique can detect the object very well.

1 INTRODUCTION

Robotics technology at this time has become very advanced. The robot is developed to have special abilities that allow it to think and behave like a human. One of the most popular robots today has the ability to play soccer. The task of this robot is to play the ball based on predefined rules for each competition (Tema KRSBI, 2017). In accordance with the rules of the robot soccer game, every object on the game field has a specific color and size. These objects are the ball, the goal frame, the white lines, and the field itself. In this research, the object of study was the goal frame.

In image processing, color filtering is the most commonly used color processing technique. The color filtering process can produce very good filters by selecting or finding a specific color from the image. With the searchable color of an image, it is possible to continue the next process. Dahlan and Budiono (2014) have conducted previous research in image processing using yellow blob detection to detect the goal image. This requires a lengthy process, while color filtering instantly results in a single-degree, black-and-white binary image without going through the threshold calibration process.

This image processing–based goal detection system uses 2D image inputs that are imagery of the goal with the x-axis, Y. Color analysis in the introduction of digital image research performed by Gudi et al. (2013) reveals several models, such as CMY, HIS, and HSV. Mandala, Rudiawan, and Soebhakti (2016) use the Mono Vision CmuCam 3.0 Camera module with the result that the robotic capability affected the object identification process, specifically the red, green, blue (RGB) range of values. In addition to RGB, HSV can recognize the ball and the goal with different colors, and the modules used vary (Kusumanto, Tompunu, & Pambudi, 2011).

Research performed using wheeled robots always focuses on the ball, but this research was more focused on the goal. The variables to be researched included the position of the robot and the background color of the goal. This will help the robot to recognize the goal as well as maximize the use of much more precise and faster image processing methods. In this study, a goal detection system based on digital image processing was performed using a wheeled robot equipped with the Logitech Camera integrated with myRIO and LabVIEW.

2 DIGITAL IMAGE PROCESSING TECHNIQUES

Digital image processing is analysis involving a lot of visual perception. In a wider definition, digital image processing also includes all two-dimensional data. In order for the imagery to be easily represented, the image needs to be manipulated into a better-quality image. Image processing, in particular, utilizes a computer to obtain a better image (Yustria, 2012).

The initial step in digital image processing is transforming an image or images into another image using a particular technique. In general, the color intensity of the digital image is divided into three, namely RGB, grayscale, and binary.

2.1 Red, Green, Blue (RGB) image

An RGB image consists of red, green, and blue elements. Essentially, the color received by the eye (the human visual system) is the result of a combination of light with different wavelengths. Research shows that the color combinations providing the widest range of colors are red (R), Green (G), and Blue (B) (Munir, 2004).

For color imagery, the RGB model – one colored image – is a three-matrix grayscale image in the form of a matrix for red (R-layer), green (G-layer), and blue (B-layer). The R-layer is a matrix that states the degree of brightness for red (e.g., for a scale of 0–255 shades, the value 0 states dark and 255 states red). The G-layer is a matrix that states the degree of brightness for green, and the B-layer is a matrix stating the degree of brightness for blue. From this definition, a certain color can be easily be presented, namely by mixing the three basic colors of RGB.

2.2 Grayscale image

A digital grayscale image is an image with the value of each pixel being a single sample. The image displayed from this type of imagery consists of gray, varying from black at the weakest part to white at the strongest intensity. Grayscale imagery differs from "black-and-white" imagery, in the context of the black-and-white computer image consisting of only two colors, "black" and "white." Grayscale imagery is stored in an eight-bit format for each pixel sample, allowing as much as 256 intensity. The lowest intensity value represents black and the highest intensity value represents white.

The image input utilized in this research was the RGB image of the goal object, so the initial step of image processing was to detect and convert the RGB image into a grayscale image. To convert an RGB image into a grayscale image, Equation (1) can be used.

$$Grayscale = \frac{R_i + G_i + B_i}{3} \tag{1}$$

2.3 Binary image

A binary image is an image that has undergone the process of separating pixels based on the strength of the gray color. Binary image formation requires a gray boundary value that will be used as the reference value. Pixels with a gray degree greater than the limit value will be rated 1 and pixels with a degree smaller than the limit value will be rated 0. This process aims to transform a gray image into a binary or black-and-white image so that the area that includes the object and background of the image is clear. The input for the threshold process is a grayscale image. The output is a binary image. The binary image can facilitate the next step in analyzing the goal frame image. The LabVIEW background is red and the goal object is black.

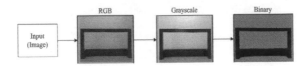

Figure 1. Digital image processing system block diagram.

3 PROPOSED METHOD

In this research, a design was proposed for a digital image processing system for goal frame detection using LabVIEW software. The goal frame detection system can be seen in Figure 1.

Figure 1 depicts the proposed digital image processing system, including shooting obtained from a camera, and the result is an RGB image. The RGB image is then converted to a grayscale image. Once converted to a grayscale image, the image will be converted again to a binary image through the thresholding process. The thresholding process generates a binary image contrasting between the color of the goal frame object and the background.

The detection process was carried out by acquiring imagery using a Logitech camera and MyRIO, which was used as a data acquisition system module to perform camera movements on a robot in a webcam-like manner in order to capture every movement.

4 RESULTS AND DISCUSSION

Image processing is one of the features of LabVIEW Vision, used for obtaining images with certain conditions. Converting RGB imagery to grayscale imagery, converting grayscale imagery to binary imagery, and the thresholding process are part of image processing.

An image processing test was done to examine whether the image processing palette on LabVIEW is used can work. Each process employed in the test has been mentioned earlier in this article.

The first step in system testing is capturing a goal object as an RGB image. Once the goal object is obtained, the next step is to process the RGB conversion to a grayscale image using color plane extraction. Extraction in digital image processing involves only one color as the dominant color of the image to be processed. LabVIEW Vision comprises a choice of colors that can be used as the dominant color: red, green, blue, hue, saturation, luminance, value, and intensity. The dominant color used in the research was value because the next process requires the values that are in the goal image. A digital image processing system was developed in the form of LabVIEW programming; program details are presented in Figure 2.

The process of converting from a grayscale image to a binary image takes an image threshold processing operation to generate a delimiter value between the main object and the background. This processing is done by mapping the qualifying pixels so that the threshold is mapped to one desired pixel value. There are upper value or upper threshold values and lower value or lower threshold values. This value can be set from 0–255 as in Figure 3.

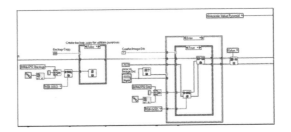

Figure 2. Digital Image Processing Program at LabVIEW.

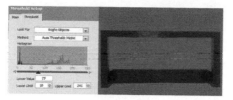

Figure 3.   Thresholding setting on grayscale image.

Figure 4.   Binary image of grayscale image through the thresholding process.

In the threshold setting, the thresholding method used was Auto Threshold – Metric. The threshold method was selected because the separation of the background and the goal object is more precise than the other methods, which still contained noise in the form of a shadow of the goal object, so that the goal object identification value had an error. The result of the thresholding process was a binary image as in Figure 4 with a red background and a black-colored goal object.

5   CONCLUSION

From the results of this discussion, it can be concluded that the object detection system based on the digital image processing technique in LabVIEW programming can be an alternative to the goal frame detection system. Because LabVIEW Vision is devoted to image processing, NI LabVIEW Vision provides specific hardware and software tools to be used for image acquisition or imagery. Choosing the thresholding method appropriately determines the identification of the destination object, because the precision of the target position is an important factor in helping the robot detect its desired object.

ACKNOWLEDGMENT

Financial support from Direktorat Riset dan Pengabdian Masyarakat, Direktorat Jenderal Penguatan Riset dan Pengembangan, Kementrian Riset Teknologi dan Pendidikan Tinggi based on the contract letter no: 040/SP2H/LT/DPRM/2019 is highly appreciated.

REFERENCES

Dahlan, A. H., & Budiono, I. 2014. Deteksi fitur dan penentuan lokasi robot pemain sepak bola berbasis penanda yang tidak unik. In *2nd Symposium on Robot Soccer Competition*.

Gudi, A. et al.2013. Feature detection and localization for the RoboCup Soccer SPL. Project Report, Universiteit van Amsterdam (February 2013).

Kusumanto, R. D., Tompunu, A. N., & Pambudi, W. S. 2011. Klasifikasi Warna Menggunakan Pengolahan Model Warna HSV. *Jurnal Ilmiah Elite Elektro, 2*(2), 83–87.

Mandala, H., Rudiawan, H., & Soebhakti, H. 2016. Sistem Deteksi Bola BerdasarkanWarna Bola dan Background Warna Lapangan pada Robot Barelang FC. *The 4th Indonesian Symposium on Robot Soccer Competition*, pp. 7–10.

Munir, R. 2004. *Pengolahan citra digital dengan pendekatan algoritmik*. Bandung: Informatika.

Tema KRSBI. 2017. *Buku Kontes Robot Sepak bola Indonesia Divisi Beroda (KRSBI Beroda) 2018*. Jakarta: Ristekdikti.

Yustria, F. 2012. Image Processing Pada Aplikasi Comic Reader Untuk Menampilkan Bagian Scene Dalam Komik Cetak.

Engineering, Information and Agricultural Technology in the
Global Digital Revolution – Hendrawan & Wijayanti Dual Arifin (eds)
© 2020 Taylor & Francis Group, London, ISBN 978-0-367-33832-9

Designing a portable LPG gas leak detection and fire protection device

Susanto, Aditya Wisnu Pradipta & Aria Hendrawan
Universitas Semarang, Semarang, Indonesia

Ma Quanjin
Universiti Malaysia Pahang, Pekan Pahang, Malaysia

ABSTRACT: Liquefied Petroleum Gas (LPG) is an alternative fuel in the form of gas that produces much fewer emissions than pollution produced by oil fuels. Therefore, on May 22, 2007, ESDM Minister Decree No: 1971/26/MEM/2007 was promulgated by implementing a program to convert fuels from kerosene to LPG. However, LPG has a greater risk of exploding than fuel oil, and many LPG cylinder explosions have occurred in Indonesia. Several security systems have been developed to address these issues, one of which is an LPG leak detection sensor. Arduino is a single, open-source microcontroller derived from a wiring-based platform. In this case, Arduino was used as a processing tool for supporting components of LPG leak detection and fire prevention devices. The author used the prototype development method. This study provided a solution to this problem by designing a gas leak detection device on an Arduino-based LPG tube using an MQ-2 sensor, as well as counteracting using a flame module equipped with a water pump in the event of a fire with a buzzer as a warning alarm and LCD for room monitoring. The detector for LPG leaks and fire prevention is expected to avoid fires caused by leaking LPG in the community.

Keywords: Arduino, Microcontroller, Robotics

1 INTRODUCTION

The Indonesian people's needs for energy have become an inseparable part of their daily lives. The community has depended on nature for the necessities of life, but the availability of natural resources – namely fossil energy, especially petroleum – is diminishing. Therefore, the use of fossil energy must be limited by switching from fossil energy to abundant natural resources, for example, natural gas energy. It has been almost 11 years since the decision of the Minister of ESDM No: 1971/26/MEM/2007, dated May 22, 2007, concerning the conversion of petroleum (kerosene) to gas (LPG) as an effort to shift from limited natural resources such as fossil energy to natural resources that are still abundant such as natural gas. Almost all people in Indonesia have switched to using LPG; besides being cheap, it is also more effective.

The LPG widely used by the community is not comparable to the gas tube producer, which has decreased in quality so that it can cause danger due to the lack of supervision of gas cylinder products. Researchers have proven in the field that many gas cylinders have been damaged, easily corroded, and dented so that they are very prone to leakage of LPG.

With current technological advances, especially in the field of information and communication technology, this issue can be handled using a microcontroller; in this case, the microcontroller used was Arduino. Arduino is an open-source microcontroller derived from the Wiring platform and designed to facilitate electronic use in various fields. The hardware has an Atmel AVR processor and the software has its own programming language. Thus the use of sensors and the right logic can be used to create a tool through which to minimize the danger of the gas cylinder. Therefore, the authors have entitled this article "Designing a Portable LPG Gas Leak Detection and Fire Protection Device."

2 THEORETICAL BASIS

2.1 *Microcontroller*

Soemarsono, Listiasri, and Kusuma (2016) contend that a microcontroller can be analogous to a computer in that all or most of its elements are packaged in one IC chip, so it is often called a single-chip microcomputer. This means that in a microcontroller IC, there is actually minimum need for a microcontroller to work, because it includes the microprocessor, ROM, RAM, I/O, and clock, as well as those of a personal computer (PC). Given the packaging is only a chip with a relatively small size, of course the specifications and capabilities possessed by the microcontroller are lower when compared to computer systems such as PCs, in terms of speed, memory capacity, and features. The difference between RAM and ROM on a computer and a microcontroller is that on a microcontroller, ROM is much larger than RAM, while in a computer, RAM is much larger than ROM.

2.2 *Arduino*

According to Syahwil (2017), Arduino is an open-source (single-source) microcontroller board derived from a wiring-based platform. This controller is designed to facilitate use in various electronic fields. Arduino's hardware contains an Atmel AVR type processor and has its own programming language. The hardware has the following specifications:

a. Fourteen (0–13) IO digital pins.
b. Digital pins numbered 0–13 can be used as input or output that is set by creating a programming IDE.
c. Six-pin (0–5) analog input.
d. Analog pins numbered 0–5 can be used to read input values that have analog values and change them to numbers between 0 and 1,023.
e. Six-pin (pins 3, 5, 6, 9, 10, and 11) analog output.

3 METHODOLOGY

3.1 *Planning phase*

Planning is the initial stage in determining the activities to be carried out by compiling and describing the aims and objectives in order to achieve the expected goals so that all activities can be directed and completed efficiently.

In building the LPG leak detection and fire protection tool using the MQ-2 and sensors flame module based on Arduino microcontrollers, of course the hardware is needed as input and output controls and software is a means for writers to write programs into the microcontroller.

What follows is a list of hardware requirements for designing LPG leak detection and fire protection devices:

a. Arduino UNO microcontroller
b. Sensor MQ-2 (as an LPG detection sensor)
c. Flame module (as a fire detection sensor)
d. 12 X 6 LCD (as a room status monitor)
e. Module Relay 5V Output 250 VAC
f. Buzzer (as a warning alarm)
g. Water pump (as a fire extinguisher when there is fire)
h. Light-emitting diode (LED)

i. Adapter 5V to 12V DC (as a microcontroller resource)
j. Resistor
k. Breadboard (experiment board)
l. Jumper
m. Cable acrylic box (as a microcontroller box)

The software needed in this design is Arduino IDE. In this case, the Arduino IDE was used as a means to write program code on the Arduino UNO microcontroller so that the range of microcontroller devices could function properly.

3.2 *Design phase*

The block diagram design was as follows:

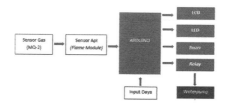

Figure 1. Block diagram of LPG leak detection and fire protection equipment using MQ-2 and sensors flame module based on Arduino microcontrollers.

Description:

a. Start the initial process.
b. Install input and output; prepare for the initial installation of input and output devices on the microcontroller.
c. Check the condition of the sensor, ensuring that the sensor is properly installed.
d. Check that the LCD indicates that the room is in a safe condition. The LCD screen displays the condition of the room as safe and turns on the green LED because the sensor does not detect any LPG or fire.
e. If the sensor detects a gas leak, it will continue to the next step in the process.
f. If the MQ-2 sensor detects an LPG leak, the LCD will read "Gas leak."
g. If the MQ-2 sensor detects an LPG leak, it will turn on the buzzer and display the red LED; if it does not, the process will revert to the LCD's indication of a safe room condition.
h. If the fire sensor (flame module) detects a fire, it will continue to the next step in the process.
i. If the fire sensor detects a fire, the LCD will display "FIRE!!!"
j. If the fire sensor detects a fire, it will turn on the buzzer and send a command to the relay to turn on the water pump and display the red LED.
k. The water pump is lit to extinguish the fire.
l. The process is complete,

4 RESULTS AND DISCUSSION

4.1 *Hardware*

Assembly hardware consisted of an Arduino UNO microcontroller, an MQ-2 sensor, a flame module, a 12 x 6 LCD display, a relay, a buzzer, an LED, a breadboard (experiment board), switches and adapters.

4.2 Assembly control

The circuit consisted of an Arduino UNO microcontroller, a relay, and an adapter that was connected with a jumper cable through pins and terminals that the author had designed previously.

4.3 Assembling the input

This input circuit consisted of MQ-2 and flame module sensors, where the MQ-2 sensor functioned as a sensor for LPG readers in the event of a leak and the flame module functioned as a reader sensor for fire. The components were connected with a jumper cable through a pin and terminal that had been designed previously.

4.4 Assembly output

The circuit consisted of a buzzer, an LED, and a water pump. Buzzers and LEDs served as alarms in the event of LPG leaks or fires. Two LEDs were utilized, namely red LEDs and green LEDs, where red LEDs indicated unsafe room conditions, while green LEDs indicated safe room conditions. A water pump combated fires, and all components were connected with a cable jumper through pins and terminals according to the previous design.

4.5 Overall series

The next step was to combine the control circuit, input circuit, and output circuit in an acrylic box. The following is a picture of the whole series.

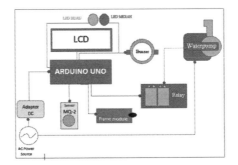

Figure 2. Prototype of LPG leak detection and fire protection equipment using MQ-2 and sensors flame module based on Arduino microcontrollers.

5 CONCLUSIONS

1. The LPG leak detection device and counter control using the MQ-2 and flame module based on Arduino microcontrollers can run well and as expected.
2. The power source of this tool was an electric current on the PLN network with an adapter; the tool will be extinguished in a power outage.
3. It is necessary to arrange a water channel installation because this tool uses a water pump as a fire extinguisher in the event of a fire.

REFERENCES

Apriyadi, S. 2013. Design of microcontroller based mobile fire detector systems. *Tanjung Pura FT University Journal*, 2(7).

Dimyati, H., & Nurjaman, K. 2014. *Project Management*. First printing. Bandung: Setia Library.

Egal. 2017. Buzzer – Arduino Uno, PCR Sites. http://egal.blog.pcr.ac.id/2017/06/10/buzzer-arduino-uno, downloaded on April 14, 2018.

Electronics2000. 2018. LED, Simon chartered. www.electronics2000.co.uk/, downloaded on November 21, 2018.

Job vacancies, et al. 2015. LPG detectors using mega-2 sensors based on mega 328 microcontrollers. *SPECTRUM E-Journal*, 2(4).

Kadir, A. 2015. *From Zero to a Pro Arduino*. Yogyakarta: Andi.

Kurnianingsih, D. 2014. Fire detection devices using Arduino R3 with LM35DZ sensor, flame sensor and MQ2. *Telematics MKOM* 6(2).

lelong.com. 2007. MQ2 MQ-2 natural gas/smoke sensor for Arduino. www.lelong.com.my/mq2-mq-2-nat ural-gas-smoke-sensor-arduino-bekind-F1203890-2007-0 1-Sale-I.htm, accessed on October 4.

Pressman, R. S. 2012. *Software Engineering*. Yogyakarta: Andi.

Ramadhan, et al. 2017. LPG gas leak detection system using fuzzy method implemented with a real time operating system (RTOS). *Journal of Information and Computer Technology Development*, 1(11).

Soemarsono, B. E., Listiasri, E., & Kusuma, G. C. 2016. Alat Pendeteksi Dini Terhadap Kebocoran Gas LPG. *TELE*, 13(1).

Studylibid. 2018. LPG (Liquified Petroleum Gas). https://studylibid.com/doc/1001980/bab-ii-landasan-teori-2.1-lpg–liquified-petroleum-gas, downloaded on October 4, 2018.

Syahwil, M. 2017. *Panduan Mudah Belajar Arduino Menggunakan Simulasi Proteus*. 1st edition. Yogya-karta: Andi.

Wikikomponen. 2018. Working principles of water pump machines. www.wikikomponen.com/pr prin ciple, downloaded on April 14, 2018.

Yendri, et al. 2017. Designing a house fire detection system for residents in urban areas based on microcontrollers. *UMJ Journal*, 2(10). components101.com, "Active-Passive Buzzer." https://compo nents101.com/buzzer-pinout-working-datasheet, downloaded on October 4, 2018.

Zonaelektro. 2018. Resistor characteristics and functions. http://zonaelektro.net/resistor-karakteristik-nilai-dan-fungsinya/, downloaded on November 21, 2018.

Engineering, Information and Agricultural Technology in the Global Digital Revolution – Hendrawan & Wijayanti Dual Arifin (eds)
© 2020 Taylor & Francis Group, London, ISBN 978-0-367-33832-9

Designing automatic system smart curtains using NodeMCU based on android

Atmoko Nugroho, Aria Hendrawan, Agusta P.R. Pinem, Saifur Rohman Cholil & Nur Hidayanti
Universitas Semarang, Semarang, Indonesia

ABSTRACT: Technology has rapidly developed in Indonesia from time to time. Technology is one important element for the development of a country, and many new technologies are available that improve the quality of human life, and that have in fact become integral to it. Sunlight also brings various benefits to human life, and curtains have an important role in controlling the amount of sunlight that enters a room. Curtains also maintain house occupants' privacy, and the utilization of technology for such household purposes is now greatly necessary. One solution is to design and implement smart curtain automation using NodeMCU. A curtain will automatically open when it detects sunlight and close in the dark. In addition, users can open and close the curtains from a distance by utilizing a Wi-Fi connection with an Android smartphone.

Keywords: NodeMCU, Smart Curtain, Android

1 INTRODUCTION

Technology has rapidly developed in Indonesia from time to time, and technology is one important element in the development of a country. Many new technologies are intended to make human life easier and they have become an integral part of our life.

Sunlight also brings various benefits for human life, including at home. However, if not properly handled, sunlight may cause discomfort. Curtains have an important role as the controller of sunlight coming into a room and in maintaining house occupants' privacy.

The use of technology for household purposes is greatly necessary. One solution is the automation of household appliances and features, and researchers have conducted research into a "Smart Curtain Automation System Design Using Android-Based NodeMCU." This smart curtain was designed to automatically open when it detects sunlight and to close in the dark. In addition, users can open and close these curtains from a distance by utilizing a Wi-Fi connection with an Android smartphone. The problem formulated here is how to design a smart curtain automation system using the Android-based NodeMCU so that it will function properly. In this study, the authors limit the problem to how the automatic curtain works, and to how the curtain driver is controlled by a microcontroller communicating with NodeMCU ESP8266, which is accessed using Android utilizing a Wi-Fi connection. The implementation of the Android system application is controlled using the Blynk application. The device made in this study is only a simulation using miniatures.

2 METHODOLOGY

In this research, the system development method used was a prototype.

1. Communication.
   Communication was made between the software development team and users. The development team held a meeting with the stakeholders to define the overall objectives of the software (Roger, 2012).

2. Quick plan and quick design modeling.

The specific needs were identified and described further in iteration. The prototype-making iteration was quickly planned and the modeling was quickly designed (Roger, 2012).

3. Prototype formation.

The authors assembled a device intended for use as a system control, such as NodeMCU with ESP8266 as the microcontroller, a light sensor, a light-dependent resistor (LDR), a DC motor driver, and a dynamo, which were then connected to the curtain replica, and began coding on Arduino IDE software.

4. System transfer and feedback.

The prototype was shown directly to consumers for evaluation and in order to get constructive input. In this case, the authors tested the system with residents of RT 03/RW 02, Tegowanu District, on Sunday, February 3, 2019, and asked for their responses through questionnaires.

3 DESIGN AND FABRICATION PROCEDURES

Previous research includes "Designing a Curtain Control with an Automatic Light Switch Based on Android Smartphone" (Ihsanto & Rifky, 2015), "Design and Implementation of a Curtain Open-Close Device Based on a Microcontroller and Android" (Wardani, Sunarya, & Ramadan, 2016), and "Designing a Vertical Blind Automation System for Curtains and Room Lights Using Light Sensors" (Asriyadi and Ciksadan, 2018).

This research was conducted in order to develop and incorporate this preexisting research. In this research, a smart curtain automation system was developed using the Android-based NodeMCU in which a light sensor is considered the automatic controller and Android's Blynk application is the manual controller. This system may be manually controlled with the Android application from a distance (without limits) as long as the system is connected to the Internet for efficiency purposes, especially for those with busy activities outside the house (Andrianto, 2017; Lubis, Novianti, & Tony, 2012; Safaat, 2012).

4 RESULTS AND DISCUSSION

Two hardware devices are needed as system controls: a light-dependent resistor (LDR) and an Android smartphone. The light sensor is designed to automatically open and close the curtain when it detects sunlight and darkness. An Android smartphone contains an application designed to order the curtain to open and close manually (Safaat, 2012). The other required hardware devices for designing the smart curtain include components and modules. The components used to build this system were the NodeMCU ESP86266, a DC motor driver, and a dynamo (Syahwil, 2017). The module used was the LDR module. The other devices used to support success in building this system included boxes, pulleys, power cables, and others.

The particular software needed in this design was the Android application. In this system, the authors used the Blynk application to control the curtain. The other software used was Arduino IDE, known as sketch, which was uploaded to the IC EFPROM contained in the NodeMCU ESP8266 itself.

The system testing comprised a series of different tests. Two methods were used to test this system: alpha testing and beta testing. Based on the beta testing results, 21.53% of respondents strongly agreed, 51.39% agreed, and 27.08% fairly agreed. Thus, it can be concluded that this device may be well accepted in communities.

5 CONCLUSION

In the automatic mode, an LDR sensor can detect intensity not only from sunlight but also from light.

This system can be manually controlled (Android) from a distance (without limits) as long as the system is connected to the Internet. The authors have tested the system with a distance of 58 kilometers from Purwodadi Square to the system located in Semarang. The curtain stability when opened and closed depends on the system's Internet connection or it can be concluded that this system can only work as long as it is connected to the Internet.

The motor can pull a curtain weight of 1 kg. The curtain can still open and close when the weight is more than 1 kg, but not perfectly. Thus, it can be concluded that the motor can work properly depending on the curtain weight. The devices needed to design this system are the ESP8266 NodeMCU as the microcontroller, a light sensor (LDR), a DC motor driver, a dynamo, the Arduino IDE software, the Blynk application, and others.

REFERENCES

Andrianto, H. 2017. *Arduino Belajar Cepat dan Pemrograman.* Bandung: Informatika.
Asriyadi, A., & Ciksadan, C. 2018. Perancangan sistem otomatisasi tirai vertical blind dan lampu ruangan dengan sensor LDR. *Rang Teknik Journal, 1*(2).
Ihsanto, E., & Rifky, M. F. 2015. Rancang bangun kendali gordeng dengan saklar lampu otomatis berbasis smartphone Android. *Jurnal Teknologi Elektro, Universitas Mercu Buana,* 28–37.
Lubis, C., Novianti, K., & Tony, T. 2012. Perancangan prototipe sistem penerangan otomatis ruangan berjendela berdasarkan intensitas cahaya. *Prosiding SNTI 2012, 9*(1),Tahun 2012.
Roger, S. P. 2012. *Rekayasa Perangkat Lunak.* Yogyakarta: Andi.
Safaat, N. 2012. *Pemrograman Aplikasi Mobile Amartphone dan Tablet PC Berbasis Android.* Bandung: Informatika.
Syahwil, M. 2017. *Panduan Mudah Belajar Arduino Menggunakan Simulasi Proteus.* 1st edition. Yogyakarta: Andi.
Wardani, D. K., Sunarya, U., & Ramadan, D. N. 2016. Desain dan implementasi alat buka tutup gorden berbasis mikrokontroller dan Android. *eProceedings of Applied Science, 2*(2).

Engineering, Information and Agricultural Technology in the
Global Digital Revolution – Hendrawan & Wijayanti Dual Arifin (eds)
© 2020 Taylor & Francis Group, London, ISBN 978-0-367-33832-9

Usability testing for multimodal transportation application

W.F. Akbar, P.W. Wirawan & E. Suharto
Universitas Diponegoro, Semarang, Indonesia

ABSTRACT: Multimodal transportation denotes an attempt to use more than one transportation mode. In previous study, an application named Travio has been developed as a tool to help find the right mode of public transportation in order to reach certain destinations in Semarang city, Indonesia. Travio application was expected to be widely used by the public. However, the application has not been tested for its usability yet. This article discusses usability testing using System Usability Scale (SUS) method for Travio application. Some respondents were selected and asked to perform specific assigned tasks. Respondents were also asked to evaluate the application using the SUS questionnaire standard. The obtained final result of the SUS score was 49.375 which meant that the Travio application required several improvements in order to increase its usability.

1 INTRODUCTION

Major cities in Indonesia provide various modes of public transportation such as Bus Rapid Transit and intra-city minibus (*angkot*). The availability of these modes could help the community in traveling. However, to go from one place to another, it usually requires more than one mode of transportation or what is known as multimodal. Users of multimodal transportation will take public transportation at a certain point and change the other transportation mode to get to the destination. Information on the use of multimodal transportation can be found using computer-based applications.

Several studies related to multimodal application development have been carried out. Zhang *et al.* developed models, for multimodal applications called ATIS (Advanced Traveler Information System) (Zhang *et al.*, 2011). Zhang *et al.* divided the transportation networks into public and private and made a model of the two networks into an extensive network. The model was then used to develop algorithms in the ATIS application. The authors also proposed a multimodal application called Travio (Yasmin, 2018). This application was used to search for public transportation that can be used in Semarang city area, including Bus Rapid Transit and city transportation (angkot).

Travio modeled transportation networks using graphs and implements the resulting model, into graph database. To use Travio, the user must enter the name of the street where he is located as well as the place name of the destination. Travio allows users to identify the nearest place at the origin and destination, so that the resulting route is expected to be more optimal. Hopefully, this application can be widely utilized by public.

The development of applications for the public needs to be measured in terms of usability so that they can be accepted and used by the public at large. Usability is one aspect of human-computer interaction, besides effectiveness, efficiency, productivity, ease-of-use, learnability, and retainability (Hartson and Pyla, 2012). One method for measuring application usability is usability testing. Usability testing is considered necessary for applications that will be used for the public. Usability testing results indicate whether or not the user can complete the work according to specific criteria, using the application.

This article is about a study of usability testing specific for Travio application. Usability testing is considered important because this application is expected to be widely used. The results of usability testing will be used to determine whether or not the user can accept the Travio application. Based on some insight within the test, it can be decided whether there is any need for improvement from the user experience to support its use. The next part of this article describes the usability testing method used, the results obtained, the interpretation of the findings and conclusion.

2   METHOD

The evaluation of Travio application using usability testing has several stages, namely selecting respondents, asking respondents to work on assignments with the application, asking respondents to fill out questionnaires, and analyzing questionnaire results. Respondents were randomly selected BRT passengers whose age ranging from 17 to 25 years. The number of respondents is 8 people. Respondents were then asked to do the task in Table 1 without being told how to do it. After completing the assignment, respondents were asked to fill out a questionnaire containing 10 declarative statements from the System Usability Scale (SUS). Each respondent was asked to fill in the values of each statement using the Likert scale 1 to 5. Scale one means strongly disagree and five means strongly agree (Hartson and Pyla, 2012). The questions was shown in Figure 1.

The values input by the respondents are used to calculate the SUS value. The SUS value is calculated using Equation 1.

$$SUS = 2,5 \, x \left[\sum_{n=1}^{5} (U_{2n-1} - 1) + (5 - U_{2n}) \right] \tag{1}$$

In Equation 1, the value given by the respondent for the odd number question (U_{2n-1}) was decreased by one while the value for even number question (U_{2n}) was used to reduce 5. The total value was summed up and then multiplied by 2.5. Then the result was compared with the value on the Sauro-Lewis Curved Grading Scale rating scale (Lewis and Sauro, 2009) in order to obtain usability testing results. The scale is shown in Table 2. The results obtained were also measured using adjective ratings (Bangor, Kortum and Miller, 2009) in order to facilitate interpreting using words.

3   RESULT

Usability testing results were obtained after the respondent performed the certain assignment given and filled out the questionnaire. Evidently, not all respondents could do the given tasks easily. Based on Table 3, it could be shown whether or not the respondents could complete the given tasks. The value of zero meant that the respondent did not complete the job while the value of one meant the respondent could complete the task successfully.

Table 4 contained the values given by each respondent for each task assigned. This table was supplemented by the SUS value calculated by Equation 1. After completed, the SUS values were summed up and divided by the number of respondents. The SUS final score yielded a score of 49.375. This result would then be used to determine what to do with the Travio application.

Table 1.   Task list for respondent.

No	Tasks
1	Open the Travio Application
2	Enter the starting location and destination location
3	See transportation modes that can be used
4	Displays information related to transportation and routes

	Strongly Disagree				Strongly Agree
1. I think I would like to use this system frequently.	O	O	O	O	O
2. I found the system unnecessarily complex.	O	O	O	O	O
3. I thought the system was easy to use.	O	O	O	O	O
4. I think I would need the support of a technical person to be able to use this system.	O	O	O	O	O
5. I found the various functions in ths system were well integrated.	O	O	O	O	O
6. I thought there was too much inconsistency in ths system.	O	O	O	O	O
7. I would imagine that most people would learn to use this system very quickly.	O	O	O	O	O
8. I found the system very cumbersome to use.	O	O	O	O	O
9. I felt very confident using this system.	O	O	O	O	O
10. I needed to learn a lot of things before I could get going with this system.	O	O	O	O	O

Figure 1. Questionnaire.

Table 2. The Sauro-Lewis curved grading scale.

Range of SUS Value	Nilai
84.1 – 100	A+
80.8 – 84.0	A
78.9 – 80.7	A-
77.2 – 78.8	B+
74.1 – 77.1	B
72.6 – 74.0	B-
71.1 – 72.5	C+
65.0 – 71.0	C
62.7 – 64.9	C-
51.7 – 62.6	D
0.0 – 51.6	F

4 DISCUSSION

It was shown in Table 3 that several things related to whether or not the respondent was able to complete the assignment. It was also shown that almost all respondents were unable to complete the second assignment. This meant that the respondent could not determine the starting point and the end point of the trip. In other words, determining the origin and destination of the trip became an essential point on the Travio application that must be corrected.

The final result of the SUS calculation yielded a value of 49.375. This meant that the usefulness of the Travio application was still low because it was worth "F" according to Table 2. If measured by adjective ratings, the Travio application still has POOR property with "Not Acceptable"

Table 3.   Results of task completion by respondents.

| Respondent | Tasks | | | |
	Task 1	Task 2	Task 3	Task 4
R1	1	0	1	1
R2	1	0	1	1
R3	1	1	1	1
R4	0	1	1	1
R5	1	0	1	1
R6	1	0	1	1
R7	1	0	1	1
R8	1	0	1	1

Table 4.   The SUS final score table.

| Respondent | Questionnaire | | | | | | | | | | |
	1	2	3	4	5	6	7	8	9	10	SUS
R1	4	1	5	3	3	1	4	1	3	2	77,5
R2	3	3	4	1	3	3	2	2	3	3	57,5
R3	4	2	3	4	4	3	2	5	3	3	45
R4	3	4	1	2	2	3	1	3	3	1	42,5
R5	3	4	4	1	3	2	3	1	3	3	62,5
R6	4	3	4	3	4	3	3	3	4	3	60
R7	3	4	2	3	2	3	2	4	1	3	32,5
R8	1	4	1	3	2	5	1	4	1	3	17,5
Final SUS score											49,375

acceptance level. This headed to the need for improvements in order to increase the usability of Travio application so that users might not experience difficulties in using the application.

5   CONCLUSION

It has been successfully done the usability testing for Travio application using the SUS method. The SUS score yielded F grade result meaning the poor adjective rating or not acceptable. One of the main causes identified in this paper was the failure of the respondent to work on the task of determining the origin and destination of the trip. This was an important note for further improvement of the Travio application, especially in terms of user interface.

REFERENCES

Bangor, A., Kortum, P. and Miller, J. (2009) 'Determining what individual SUS scores mean: Adding an adjective rating scale', *Journal of usability studies*. Usability Professionals' Association, 4(3), pp. 114–123.

Hartson, R. and Pyla, P. S. (2012) *The UX Book: Process and guidelines for ensuring a quality user experience*. Elsevier.

Lewis, J. R. and Sauro, J. (2009) 'The factor structure of the system usability scale', in *International conference on human centered design*. Springer, pp. 94–103.

Yasmin, Y. (2018) 'Aplikasi Transportasi Multimodal Traveling Semarang Berbasis Web Menggunakan Graph Database Schema', *Prosiding Seminar Nasional Ilmu Komputer (SNIK) 2018*, pp. 66–74.

Zhang, J. *et al.* (2011) 'A multimodal transport network model for advanced traveler information systems', *Procedia-Social and Behavioral Sciences*. Elsevier, 20, pp. 313–322.

Engineering, Information and Agricultural Technology in the
Global Digital Revolution – Hendrawan & Wijayanti Dual Arifin (eds)
© 2020 Taylor & Francis Group, London, ISBN 978-0-367-33832-9

Development of airports from the passenger demand side: A case study of Dewadaru Airport, Karimunjawa, Jepara city, Central Java, Indonesia

A. Muldiyanto, M. Handajani & Supoyo
Universitas Semarang, Semarang, Indonesia

ABSTRACT: At Dewadaru Airport, Karimunjawa, Jepara, Central Java, Indonesia, currently the runway has a dimension of 1,200 meters by 30 meters, while a Twin Otter-type aircraft has a capacity of 6–12 passengers. To increase the airport's capacity, it is necessary to add a runway so that the airport can be used by ATR 72 aircraft with a capacity of 40–70 people. The purpose of this study was to predict the number of aircraft passengers 5–15 years to come using regression analysis. The results for 2033 were 74,954 passengers, and the projection results were 71,524 passengers. From the results of the prediction of the number of passengers, 2,028 ATR 72 aircraft are needed, which need a runway of 1,600 meters.

Keywords: Runway, Regression, Projection, ATR 72

1 BACKGROUND

Dewadaru Airport is located on Karimunjawa Island, Jepara, at coordinates 050 48 '04 "S - 1100 40' E and 7 meters Dpl elevation. Dewadaru has a runway measuring 1,200 meters by 30 meters. Karimunjawa Island and its surroundings boast many beautiful nautical tourism objects, so the island attracts both domestic and foreign tourists. The airport as a part of the island's transportation infrastructure certainly needs development.

Dewadaru Airport currently serves scheduled flights of pioneer air transport by PT Airfast Indonesia with Twin Otter aircraft and charter flights by Kura Kura Resort, whose passenger capacity is very limited, ranging from 6 to 12 passengers.

In addition to traveling by air to Karimunjawa Island, tourists can travel by sea, that is by Pelni and fast boats, but during the busy season, ships cannot operate because they cannot dock at the Karimunjawa port, leading to longer travel times and high unexpected costs.

Dewadaru Airport needs to be developed so that it can be used for larger aircraft types such as the ATR 72, with capacities ranging from 40–70 passengers; for this reason, there is a need to study the number of aircraft passenger requests in the future.

1.1 *Research purpose*

The purpose of this study was to estimate the number of aircraft passengers for the next 15 years using regression analysis, and to analyze the necessary airport infrastructure so that the runway can be used to serve ATR 72 aircraft.

1.2 Limitation of problems

Historical data on aircraft and tourists were from the 2013–2018 period. For regression analysis, the independent variable was the number of tourists on Karimunjawa Island, and the fixed variable was the number of aircraft passengers.

2 LITERATURE REVIEW

Airports, according to Utami (2012), are areas of land and/or water with certain limits that are used as a place for aircraft to land and take off, to board passengers, to load and unload goods, and as inter-transportation modes, that are equipped with flight safety and security facilities. Airports are important nodes in the transportation system and can play an important role in supporting the socioeconomic development of urban areas (Ferrulli, 2016). For Esturt (2016), as the current development of the airport platform changes aeronautical activities, the relationship between commercial and aeronautical activities also changes (Lopez, 2016).

Prediction is an important stage in airport planning and management, and can be used as a tool to determine runway length and when it is needed. Predictions can also be wrong because of uncertainty, so they involve a process of converting uncertainty into measurable risks (Mubarak, 2015).

Wardhani Satono and Rahman (2015) list three methods for airport planning: time series methods (time series), market reach methods (market share), and econometric methods (Malavolti, 2016). This study employed an econometric model because it is considered the best and the most complex in forecasting passenger aircraft demand. Trip generation in this forecasting uses simple and plural regression analysis, both linear and nonlinear (Mubarak, 2015).

3 METHODOLOGY

This study used an econometric model with regression analysis.

The stages in predicting the number of passengers at Dewadaru Airport were: first, make a regression equation from historical data, with as the fixed variable and the number of visitors on Karimunjawa Island as the independent variable. Second, project the number of passengers according to the percentage of growth, and predict the number of passengers using regression analysis. Third, analyze the number of daily passengers based on the prediction results. Fourth, analyze the runway's length, as it will need to accommodate ATR 72 aircraft.

4 RESULT

Historical data on aircraft passengers and the number of tourists visiting Karimunjawa in 2013–2018 can be seen in Table 1.

Table 1. Historical airline passenger and tourist data for 2013–2018.

Year	Aircraft passengers	Tourists
2013	2,076	39,224
2014	2,101	59,621
2015	3,574	81,235
2016	4,769	92,115
2017	4,992	103,169
2018	4,601	122,876

Source: Jepara in numbers, 2019

Historical data were then analyzed by regression to get the equation $Y = 0.040 X + 389,617$, where Y is the number of aircraft passengers who ride from Dewadaru Airport and X is the number of tourists visiting Karimunjawa Island, with df = 5, R = 0.900, R2 = 0.811. The linear regression significance test revealed that 81.1% of passenger departure variations can be represented by this equation and F = 17,140 > 6.61 (table) and t = 4.14 > 2.015 (table) for the degree of freedom 5 and 95% confidence level. The X coefficient value is tourists visiting Karimunjawa.

The projection of the numbers of aircraft passengers and travelers up to 2033 was based on the average growth, where the growth of aircraft passengers and tourists, respectively, were 12.69% and 19.88%; the results can be seen in Table 2.

Comparisons of the results of the five-year aircraft passenger projections from the regression equation up to 2033 can be seen in Table 3 and Figure 1.

The results of the prediction of the number of aircraft passengers are depicted in Table 3. The average number of flight passengers can be calculated, and those results are seen in Table 4.

From the results shown in Table 4, in 2028, the number of daily airplane passengers is projected to be 85, meaning that before 2028, aircraft passengers can be adequately served by increasing the frequency of flights with existing aircraft. After 2028, it may become necessary to extend the runway in order to use the ATR 72 type aircraft.

Table 2. Airplane passenger and tourist projections until the year 2033.

Year	Aircraft passengers	Tourists
2019	5,185	147,299
2023	8,360	304,183
2028	15,190	753,013
2033	27,600	1,864,103

Source: Calculation results

Table 3. Projection results and passenger predictions.

Year	Projections	Predictions
2019	5,185	6,282
2023	8,360	12,557
2028	15,190	30,510
2033	2,760	74,951

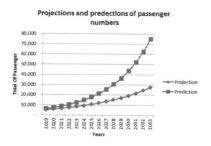

Figure 1. Projection graph and prediction of number of passengers.

Table 4. Average daily passenger predictions.

Year	Daily aircraft passengers
2019	17
2023	35
2028	85
2033	208

Source: Calculation results

Runway length assumptions can be calculated using ATR 72-600 ARFL = 1,333 m aircraft types. Existing data (source: Dewadaru Airport) include: elevation = 7 MSL, average temperature = 280°C, and slope = 0.167. These data can be used with the ICAO method:

$$ARFL = \frac{Actual\ Runway\ Length}{Fe\ x\ Ft\ x\ Fs}$$

Elevation correction factor = Fe = 1 + (0.07x (e/300) = 1 + (0.07 x (7/300), Fe = 1.0016; Temperature correction factor = Ft = 1 + (0.01 x (T− (15−0.0065x e) = 1+ (0.01x (28− (15−0.0065x7), Ft = 1.1304; Correction factor slope = 1+ (0.1 x S) = 1 + (0.1 x 0.167), Fs = 1.0167. So the total F = Fe x Ft x Fs = 1.0016 x 1.1304 x 1.0167; F = 1.1512.

Actual runway length = ARFL x F = 1,333 x 1.1512 = 1,535.62 m rounded to 1,600 m.

The runway width is still 30 meters, so that later it will include 3C class airports, according to ICAO Annex 14, Aerodromes, Volume 1, *Aerodrome Design and Operation* (ICAO Annex 14, 2016).

5 DISCUSSION

Karimunjawa Island and its surroundings boast many beautiful nautical tourism objects, so the island attracts both domestic and foreign tourists. The airport as a part of the transportation infrastructure certainly needs development. One factor driving airport development is passenger demand, and predictions can made for the next few years, from the regression equation Y = 0.040X + 389,617. This equation can be used to calculate the number of aircraft passenger requests at Dewadaru Airport for the next few years. From the passenger requests, it can be known whether the airport needs to be developed or if merely increasing the number of flights will meet the demand. The real condition in the field is that the ATR 72 aircraft, starting in April 2019, has been tested in both direction along the Semarang–Karimunjawa route. In fact, the ATR 72 aircraft is currently only filled with 40 passengers, which should be 70 passengers, because there is insufficient runway space on which the aircraft can land when filled to capacity.

6 CONCLUSION

From the regression equation with the prediction presented earlier in this article, it can be concluded that it is necessary to increase the number of flights to and from Dewadaru Airport in order to overcome the growth in the number of aircraft passengers until 2028; after 2028, the runway must be extended by 1,600 meters, assuming an ATR 72 aircraft is deemed necessary after further study and more technical airport planning.

REFERENCES

Ferrulli, P. 2016. Green Airport Design Evaluation (GrADE): Methods and tools for improving infrastructure planning. *Transportation Research Procedia, 14*, 3781–3790.

ICAO Annex 14. 2016. *Aerodrome Design and Operations*. 7th edition.

Lopez, A. 2016. Vulnerability of airports to climate change: An assessment methodology. *Transportation Research Procedia, 14*, 24–31.

Malavolti, E. 2016. Single till or dual till at airports: A two-sided market analysis. *Transportation Research Procedia, 14*, 3696–3703.

Mubarak, T. 2015. Airport passenger demand forecasting using radial basis function neural networks: The Juanda International Airport case. Yogyakarta: Gadjah Mada University.

Utami, A. H. D. 2012. *Analisis Pengembangan Runway Dan Fasilitas Alat Bantu Pendaratan Di Bandar Udara Depati Amir Bangka*. Skripsi: Jurusan Teknik Penerbangan Sekolah Tinggi Teknologi Adisutjipto Yogyakarta.

Wardhani Satono, D., & Rahman, T. 2015. *Bandar Udara: Cetakan 1*. Yogyakarta: Gajah Mada University.

*Engineering, Information and Agricultural Technology in the
Global Digital Revolution – Hendrawan & Wijayanti Dual Arifin (eds)*
© 2020 Taylor & Francis Group, London, ISBN 978-0-367-33832-9

Area-based segmentation in brain scan CT images for low-grade glioma patients

Aria Hendrawan, Atmoko Nugroho, Saifur Rohman Cholil & Agusta Praba Pinem
Universitas Semarang, Semarang, Indonesia

ABSTRACT: Glioma is a primary brain tumor disease that often attacks adults. Gliomas are related to nerve cells and other tissue infiltrations. Based on the level of spread in the brain, this disease is divided into high-grade and low-grade glioma. Much research has discussed this disease, but research related to early detection of glioma remains limited. In this study, we developed a segregation method for an area-based CT scan of a brain with low-grade glioma. The initial process needed to produce area-based image segmentation was preprocessing using a threshold, then the image segmentation process used fuzzy classification with three entropies. We consulted a neurologist to get information about the CT scan of the area of the brain with glioma. The final result of this study was segmentation of the CT scan. The results of this segmentation can be used by specialists to detect early glioma.

Keywords: Glioma, Low-Grade Glioma, Image Segmentation, Fuzzy

1 INTRODUCTION

Glioma is a primary brain tumor disease that generally attacks adults and that may originate from glial cells (tissue that binds cell nerves and fibers) and infiltration of surrounding tissues. Glioma is usually associated with nerves in the spine. Although research on glioma has made great progress, early detection in patients with glioma is very limited. Early detection of glioma is very important because patients need immediate care; a patient suffering from high-grade glioma has an average life expectancy of only up to two years. To find glioma, a brain image is needed. A head CT scan is a method of radiological examination used to evaluate patients who have a head injury.

The results of CT scans of the head, especially in the brain, can be used to diagnose glioma. Tomographic analysis performed by a specialist in relation to glioma will determine the location of the glioma in the patient's brain. Through tomographic analysis of CT scan data, specialists can diagnose the level of glioma (high- or low-grade glioma) and take appropriate medical action. The diagnosis by a specialist who reads a CT scan image is subjective, which means that each doctor will have a different diagnosis depending on the clarity of the CT scan image he sees.

Segmentation of CT scan images is needed to clarify some images for diagnosis. Through the CT scan segmentation method, especially for detecting glioma, the area of the brain indicated by glioma can be accurately known. The segmentation method can achieve clearer CT scan images for areas of the brain indicated by glioma so that specialists can objectively diagnose glioma. For these reasons, the authors intend to develop a method of segmentation for CT scan brain images in patients with glioma.

2 METHODOLOGY

The object of research in this study was the brain image of low-grade glioma sufferers; the image was produced through CT scans. The method of data collection in this study consisted of primary data collection methods conducted by the author via collecting medical images taken from the results of CT scans. Medical images were obtained from public datasets originating from publicly available cancer-imaging websites that have URL addresses. The data used were data that followed a Digital Imaging and Communication in Medicine (DICOM) format, totaling 20 pieces of data. Secondary data collection methods included literature studies of journal articles relating to the segmentation of CT scan images for glioma.

The flow of this research appears as follows:

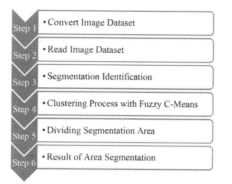

Figure 1.   Flow of research on image segmentation of brain CT scan.

3 RESULT AND DISCUSSION

With MicroDicom software, files that have the DICOM format were exported to JPG format files. In the process of reading images, converting DICOM images into JPG images resulted in color images. These color images needed to be converted into a gray image (grayscale) so that the segmentation process would go more quickly. In this process, each image area with similarities in pixel intensity and characteristics were identified and grouped in a particular cluster. In this clustering process, the grouped areas, especially the images of the areas of the brain with glioma, were bounded and determined by the clustering center. The clustering center was determined by four values: namely in the matrix that has membership values of $cc1 = 8$, $cc2 = 90$, $cc3 = 180$, and $cc4 = 250$. Then the cluster center was calculated by iterations. The objective function was used as a recurrence requirement to get the right cluster center. So the tendency of the data to enter the same cluster was obtained in the final step.

To get more accurate results, we consulted a neurologist. In Figure 2 can be seen the result of segmentation – the blue area is the area of the brain with glioma.

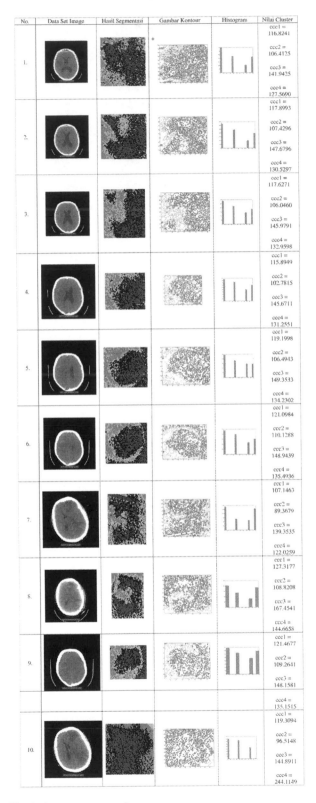

Figure 2.   Results of brain image segmentation.

11.				ccc1 = 140.4609 ccc2 = 113.1264 ccc3 = 175.9356 ccc4 = 241.1801
12.				ccc1 = 128.0351 ccc2 = 111.9196 ccc3 = 153.0782 ccc4 = 140.7791
13.				ccc1 = 119.6013 ccc2 = 99.4373 ccc3 = 144.4414 ccc4 = 235.2441
14.				ccc1 = 145.3192 ccc2 = 117.5914 ccc3 = 178.6853 ccc4 = 245.4994
15.				ccc1 = 124.2101 ccc2 = 101.6119 ccc3 = 157.1330 ccc4 = 249.8252
16.				ccc1 = 147.3471 ccc2 = 119.7016 ccc3 = 183.0914 ccc4 = 248.6154
17.				ccc1 = 122.1546 ccc2 = 100.2933 ccc3 = 157.1338 ccc4 = 250.6288
18.				ccc1 = 146.2125 ccc2 = 119.4999 ccc3 = 181.3210 ccc4 = 246.3979
19.				ccc1 = 75.4456 ccc2 = 88.5521 ccc3 = 115.8041 ccc4 = 101.1154
20.				ccc1 = 75.6699 ccc2 = 88.9787 ccc3 = 120.0390 ccc4 = 102.4196

4 CONCLUSION

Area segmentation carried out in this study succeeded in dividing the CT scan of the brain into three areas – namely, the yellow, dark blue, and light blue areas. The dark blue area depicts the part of the brain with glioma.

ACKNOWLEDGMENT

Our thanks go to Dr. A. Gunawan Santoso, Sp.Rad (K) who helped us in reading the results of our segmentation of brain CT images.

REFERENCES

Goel, S., Sehgal, A., Mangipudi, P., & Mehra, A. 2016. Brain tumor segmentation in glioma images using multimodal MR imagery. *Proceedings of the International Conference on Intelligent Communication, Control and Devices* (September 18), 733–739.

Gupta, D., Anand Hospital, & Tyagi, B. .2011. Feature extraction and identification of MRI based on brain image. *Technical Forum, 34*(1).

Gupta, D., Anand Hospital, & Tyagi, B. 2015. A hybrid segmentation method based on Gaussian fuzzy clustering: Kernel and region-based active model for ultrasound medical images. *Biomedical Signal Processing and Control, 16*, 98–112.

Huang, M., Yang, W., Yao, W., Jiang, J., Chen, W., & Feng, Q. 2014. Brain tumor segmentation based on local independent projection classification. *IEEE Transactions on Biomedical Engineering, 61*(10).

Li, Z., Wang, Y., Yu, J., Shi, Z., Guo, Y., Chen, L., & Mao, Y. 2017. Low-grade segmentation glioma based on CNN with fully connected CRF. *Journal of Healthcare Engineering.*

Liu, J., Li, M., Wang, J., Wu, F., Liu, T., & Pan, Y. 2014. A survey of MRI-based brain tumor segmentation methods. *Tsinghua Science and Technology,* 19(6),578–595.

Sun, J., Wang, Y., Xiaohong, X., Zhang, X., & Gao, H. 2012. New image segmentation algorithm and its application in lettuce object segmentation. TELKOMNIKA, 10(3),557–563.

*Engineering, Information and Agricultural Technology in the
Global Digital Revolution – Hendrawan & Wijayanti Dual Arifin (eds)
© 2020 Taylor & Francis Group, London, ISBN 978-0-367-33832-9*

Effect of mixed dimethyl ether in LPG to flame stability characteristic jet diffusion flame

L. Aulia & R Anggarani
Department of Mechanical Engineering, University of Indonesia, Depok, Indonesia
PPPTMGB 'LEMIGAS', Cipulir, Jakarta Selatan, Indonesia

E. Yuliarita
PPPTMGB 'LEMIGAS', Cipulir, Jakarta Selatan, Indonesia

Susanto
Department of Mechanical Engineering, University of Indonesia, Depok, Indonesia
Faculty of Engineering, Universitas Samudra, Meurandeh, Langsa, Indonesia

I Made K Dhiputra
Department of Mechanical Engineering, University of Indonesia, Depok, Indonesia

ABSTRACT: Utilization of LPG (Liquefied Petroleum Gas) as fuel for the agricultural products processing sector and the food industry is in demand because LPG has the characteristics of combustion gases that are cleaner than liquid fuel or other conventional fuels. Until now, the government imported 70% of domestic LPG needs, mostly to meet the industrial and household sectors. DME (Dimethyl Ether) is an alternative fuel that has considerable potential as a substitute for LPG. In addition to being characterized by having fuel characteristics similar to LPG, it can also be produced from biomass and coal, which have large reserves in Indonesia. For initial use, DME can be used as an LPG mixture. In this study an experimental study was conducted on the effect of DME as a mixture of LPG gas on the flame stability characteristics. With variations in the mixture of 10%, 20%, 30%, 40% and 50% by weight, the results of the change in speed stability properties of blow off velocity, lift off velocity and lift off height from the LPG mix DME jet diffusion flame are obtained. In addition, the DME mixture effect shows an increase in the combustion speed of the DME LPG mix

1 INTRODUCTION

The use of LPG (Liquefied Petroleum Gas) as fuel for the processing sector of agricultural products and food processing has been widely used in Indonesia. This is because LPG combustion emissions are cleaner than other conventional fuels such as liquid fuel and biomass. Now, the government imported 70% of domestic LPG needs, mostly to meet the industrial and household sectors. DME (Dimethyl Ether) is an alternative fuel that has considerable potential as a substitute for LPG. Beside having fuel characteristics similar to LPG, it can also be produced from biomass, coal and waste (methane). DME is the simplest ether compound (CH_3-O-CH_3), in the form of gas in ambient temperature, and can be liquefied like LPG so that the infrastructure for LPG can be used for DME. In its application as fuel, DME has similar characteristics to LPG. The use of DME will produce a low environmental impact, where the combustion does not produce sulfuric acid (SOx) and smoke, and produces very low NOx and CO_2 Boedoyo 2016. DME is non-toxic, easily liquefied and easy in handling, the DME Application can cover several sectors, including: automotive, domestic (household), power generation, and industrial raw materials chemistry.

Therefore, Marchionna *et al.* (2008); Anggarani, Wibowo and Rulianto (2014) explained that the application of DME fuel mixed with LPG between 15-20% can be done without modification in the household stove system. Xue and Ju (2006) states that DME has a different phenomenon of lift-off characteristics than other hydrocarbon fuels in jet diffusion flame. Lee, Sung and Park (2012) explained in his research about combustion characteristics and NOx emissions of LPG fuel, DME and its mixtures using diffuse counter flow flame burners that with higher amounts of DME mixtures in LPG causing flame thickness increases, but the length of the flame decreases. Zhou, Ai and Kong (2013) found that by preheating the DME, the lift-off nature of DME could occur by increasing the amount of mass of the fuel flow (jet velocity). Kang *et al.* (2015) conducted observations on the characteristics of stability behavior, the volume, width and length of the fire and flow and mixing using DME, methane and LPG jet diffusion flame fuels. The test uses 5 pieces of nozzles with different diameters and variations in the jet flow velocity, starting from the laminar pattern to the blow-out and producing a form that has no dimension to the observations. The use of dryers in maintaining and improving the quality of post-harvest agricultural products is important. This study aims to mix LPG fuel with Dimethyl Ether (DME) to determine the effect on flame stability compared to LPG.

2 EXPERIMENTAL

In this study used LPG (propane-butane mixture) produced by PT. PERTAMINA (Persero) and DME with 99.9% purity produced by PT. Bumi Tangerang Gas Industry. DME LPG mix fuel used in this study varied the mixture starting from DME 10% LPG mix (LMD10), DME 20% LPG mix (LMD20), DME 30% LPG mix (LMD30), DME 40% LPG mix (LMD40) and DME 50% LPG mix (LMD 50). The Blending Process of DME fuel in LPG is carried out in a combustion laboratory research group Fuel Research Center and Development of oil and gas "LEMIGAS". In Table 1, the characteristics of DME LPG mix fuel are displayed as a result of blending.

In Figure 1. the flame stability testing scheme is shown in this study. Testing is done by varying the fuel flow rate, then each flame test is documented using a digital camera.

3 RESULTS AND DISCUSSIONS

Flame stability is usually seen from lift-off velocity, lift-off height and blow-off velocity. Lift-off velocity is the speed of the fuel flow rate where the flame is lifted from the burner nozzle. This happens if the velocity of the fuel flow is greater than the velocity of the local laminar flame. Smaller the velocity of flowrate, the distance between the flame base and the burner nozzle is more closer, and so the opposite. Lift-off height indicates the distance between the burner nozzle and flame base. Blow off is a condition where the flame extinct due to the flow velocity limit greater than the burning velocity. In Figure 2 shown flame height at the variation velocity of fuel flowrate. The picture is used as a visual comparison of the high-difference LPG flame, DME and DME LPG mixed. In the picture can be seen that there is a difference in height at velocity variation of the fuel flow rate. The difference in the amount of fuel flow rate for LPG, DME and DME LPG mix is due to the blow-off phenomenon that occurs for each fuels. In LPG fuel blow-off occurs at 28 SCFH flow rate (13.1 ltr/minute), while in DME fuel blow-off occurs at 15 SCFH flow rate (7.08 ltr/minute) and on LPG mixed 50% DME occurs blow-off at a flow rate of 20 SCFH (9.4 ltr/minute). The rotameter used has a scale of 3 SCFH per line and the maximum measurement capability is up to 30 SCFH. In the flow rate of 3 SCFH (1.4 ltr/min), 6 SCFH (2.8 ltr/min) and 9 SCFH (4.2 ltr/min) there is a significant increase in flame height, but starting at 9 SCFH, 12 SCFH (5.6 ltr/min), 15 SCFH (7.08 ltr/min) and 18 SCFH (8.4 ltr/min) the increase in the height of the flame is very small and almost the same. This is due to the change from the laminar combustion to turbulent along with the increase in velocity flow rate of the fuel so that the phenomenon of lift-off occurs on the jet diffusion flame so that the flame height does not increase.

Table 1. Physical properties of LPG mix DME.

Characteristics		Unit				DME (%) Limits		
	LMD10	LMD20	LMD30	LMD40	LMD50		Min	Max
Specific Gravity @60/60°F	-			0.5468	0.5602	0.5703	0.5839	0.5976
to be reported		Vapor Pressure @100°F			Psig			102
103		104		105	107			-
145								
Weathering Test @36°F		% vol		99.5	99.6		99.6	
99.7	99.7			95	-			
Copper Strip Corrosion	-			1b	1b			1b
1b			1b		1b			
ASTM No.1								
Total Sulphur				Grains	1.89			1.75
1.61		1.55		1.43	-			15
Water Content		-					No FW	No
FW	No FW	No FW	No FW	No FW	-			
Composition:								
DME					% vol			11.41
22.73		32.23		42.56	52.84			-
-								
C2					% vol			0.15
0.11		0.09		0.08	0.06			-
0.8								
C3 dan C4					% vol		88.17	
76.9		67.35		57.11	46.97		97.0	-
C5+ (heavier)					% vol	0.27		0.26
0.33		0.25		0.13	-			2.0

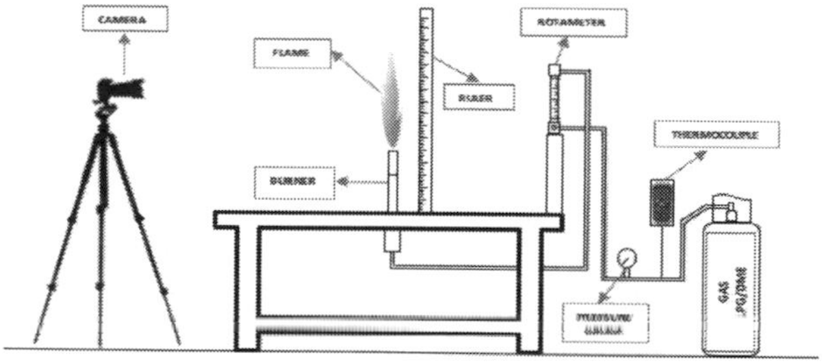

Figure 1. Jet diffusion flame testing scheme

In Figure 2, it is shown the characteristic of flame stability on the speed of the mass flow rate of the fuel. The results of the tests show that DME has a greater combustion reaction rate than LPG.

3.1 Blow off and lift off phenomena

In Figure 3, the graph shows the effect of mixing DME in LPG on the flame blow-off and lift-off. In the graph it can be seen that the greater the addition of DME concentration in LPG gas accelerates the blow-off with a lower fuel flow rate. The blow-off phenomenon in fuel

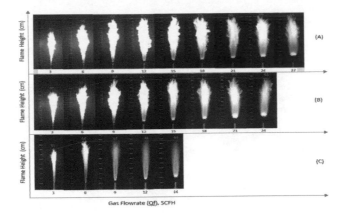

Figure 2.   Flame Height of LPG (A); LPG Mix DME 50% (B); DME (C).

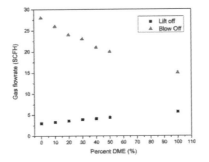

Figure 3.   Lift Off and Blow off Flame Characteristics of LPG mix DME.

without DME mixture occurs when the fuel flow rate is 28 SCFH (13.2 liters/minute), while for pure DME occurs when the fuel flow rate is 15 SCFH (7.08 liters/minute). This proves that DME fuel has a higher combustion speed compared to LPG fuel. In flame jet diffusion, the blow-off phenomenon is followed by the occurrence of the lift-off phenomenon or the rising flame. Increasing the velocity flowrate of the fuel for blow-off triggers an increase in the distance of the flame from the burner nozzle. Beside measure flame dimension of jet diffusion using a mixture of LPG, DME and DME LPG mixed, we also measured the velocity flowrate of fuel in the lift-off on a jet diffusion flame. And Figure 3 also shows the effect of mixing DME in LPG gas on the lift-off phenomenon when the fuel flow rate increases. The graph shows that there is a difference in the fuel flow rate for lift-off. At the fuel mass flow rate of 3.3 SCFH (1.41 ltr/minute) LPG gas starts to rise, but at DME fuel the flame starts to rise at a fuel flow rate of 5.5 SCFH (2.6 ltr/minute) This is a unique phenomenon where LPG and DME have similar physical chemical characteristics in addition to the combustion rate of DME greater than LPG. The test results also show that an increase in the amount of DME mixture in LPG gas (DMG LPG mixture) causes changes in the nature of LPG fuel removal.

4   CONCLUSION

Based on the experiment results, we can conclusion that there is a significant effect of mixing DME in LPG to flame stabilizations of jet diffusion flame. DME influence the blow off velocity on LPG mix DME to be rise significantly. The DME effect also has a different impact on lift off velocity. Increasing the fuel flow rate velocity is not followed by the magnitude of the life off velocity. The burning velocity also increases. This is caused by the presence of Oxygen content in DME molecules.

REFERENCES

Anggarani, R., Wibowo, C. S. and Rulianto, D. (2014) 'Application of dimethyl ether as LPG substitution for household stove', *Energy Procedia*. Elsevier, 47, pp. 227–234.

Kang, Y.-H. *et al.* (2015) 'Experimental and theoretical study on the flow, mixing, and combustion characteristics of dimethyl ether, methane, and LPG jet diffusion flames', *Fuel processing technology*. Elsevier, 129, pp. 98–112.

Lee, T. S., Sung, J. Y. and Park, D. J. (2012) 'Experimental investigations on the deflagration explosion characteristics of different DME–LPG mixtures', *Fire Safety Journal*. Elsevier, 49, pp. 62–66.

Marchionna, M. *et al.* (2008) 'Fundamental investigations on di-methyl ether (DME) as LPG substitute or make-up for domestic uses', *Fuel Processing Technology*. Elsevier, 89(12), pp. 1255–1261.

Xue, Y. and Ju, Y. (2006) 'Studies on the liftoff properties of dimethyl ether jet diffusion flames', *Combustion science and technology*. Taylor & Francis, 178(12), pp. 2219–2247.

Zhou, J., Ai, Y. and Kong, W. (2013) 'The Liftoff Properties of Dimethyl Ether Jet Diffusion Flames With Preheating', in *ASME Turbo Expo 2013: Turbine Technical Conference and Exposition*. American Society of Mechanical Engineers Digital Collection.

*Engineering, Information and Agricultural Technology in the
Global Digital Revolution – Hendrawan & Wijayanti Dual Arifin (eds)
© 2020 Taylor & Francis Group, London, ISBN 978-0-367-33832-9*

Study of the role of CFRP shear on increased bending capacity of reinforced concrete beams

Hari Suprapto & Sri Tudjono
Universitas Diponegoro, Semarang, Indonesia

Rr. MI. Retno Susilorini
Senior Lecturer, Civil Engineering Department, Soegijapranata Catholic University, Semarang, Indonesia

ABSTRACT: In order to restore structural strength, improvements can be made so that the structure can function as expected. This study aimed to examine the role of shear Carbon Fiber-Reinforced Polymer (CFRP) confinement in the flexural strength of reinforced concrete beams. The beam was designed as an over-reinforced beam. Observations focused on pure bending spans, which meant that failure did not occur in the shear area. A normal beam without CFRP treatment served as a control using the same material and loading specifications. This study indicated that the CFRP restraint can increase the bending moment capacity of the beam by 32.38%. Ductility, as an indicator of the ability of a structural element to improve, also increased, by 43.14%.

Keywords: CFRP, Restraint, Over-Reinforced, Bending Capacity, Ductility

1 INTRODUCTION

Strength degradation in buildings can be caused by factors such as detriment, fatigue, age, and deficiencies in construction or planning. Building strength can also change when there is a transfer of functions to the building. Carbon fiber-reinforced polymer (CFRP), which is generally in the form of sheets, can be affixed to parts of structural elements that require repair (Sobuz et al., 2012). The CFRP materials can be used to increase structural strength and can also be applied to damaged structures. The use of CFRP has been regulated in ACI-440 (ACI 440 2R-08, 2008), both as flexible and as shear reinforcement. This research examined the effect of CFRP restraints on increasing flexural strength in reinforced concrete beams (Park & Paulay, 1974).

2 REVIEW OF PREVIOUS STUDIES

Researchers have studied CFRP as a reinforcement material used for the improvement of reinforced concrete beam structural elements. Hadi (2009) examined four beam specimens consisting of a normal B beam, a BF beam with three CFRP layers without fiber steel, a BS beam with an additional 1% steel fiber without CFRP, and BFS beams with an additional 1% steel fiber and three layers of CFRP. The four beams were loaded with loads centered on four locations on the span. The test results showed that CFRP can increase both load capacity and ductility. Pangestuti and Handayani (2009), in their work on the role of CFRP in single reinforced concrete beams, showed that there was an increase in stiffness of 33.3% and an increase in cracking moments of up to 50% compared to the same beam that was not treated with CFRP. Research by Utami, Nuroji, and Antonius (2016) studied the use of CFRP by means of dressing on reinforced concrete beams with a cross-section size of 15 cm x 25 cm and a span length of 200 cm. The test results showed that CFRP band usage beams can increase ultimate load capacity from normal conditions, which are between 0.84% and 14.88%. The test results also showed that using CFRP can increase the beam displacement ductility by between 33% and 45%. Hidayat's (2016) study on reversed T-beams showed that reinforced concrete beams treated with CFRP experienced an increase in capacity of up to 51.92% compared to untreated beams.

From the various studies in general, it can be concluded that the use of CFRP materials plays a significant role in increasing the capacity of reinforced concrete beams.

2.1 *Theoretical basis*

The nominal moment strength of a square beam can be calculated according to the equation:

$$M_{n0} = \left(A_s - A_s'\right)f_y\left(d - \frac{a}{2}\right) + A_s'f_y\left(d - d'\right) \tag{1}$$

ACI 440 article 10 states that the nominal flexural strength of a square cross-section strengthened by CFRP is formulated as:

$$M_{nf} = A_s f_s\left(d - \frac{\beta_1 c}{2}\right) + \psi_f A_f f_{fe}\left(h - \frac{\beta_1 c}{2}\right) \tag{2}$$

In this study, CFRP was linked only to depressed areas along pure flexure spans. Increased load was carried out until a crack appeared, and so on until a press failure occurred. The next degree of flexural strength change was calculated as the relative difference between the flexural strength of the treated beam and the normal beam.

Ductility is the ability of a structure to deform significantly, without experiencing a significant decrease in stiffness. This study reviewed the displacement ductility, which is a comparison between the total vertical displacement value that occurs when the beam has reached the point of collapse and a vertical displacement value at the time of the first yielding.

3 THE METHOD

The model comprised a section that measured 15 cm x 30 cm x 240 cm. Concrete material was designed with the quality of fc' = 18.7 MPa; for tension reinforcing steel 6D16, the quality was fy = 390 MPa; the stirrup was ϕ8–50 mm with quality fy = 240 MPa. The CFRP material specifications follow: ultimate pull strength f*fu = 4,900 N/mm²; ultimate fraction ε*fu = 0.015; elasticity Ef = 230,000 N/mm²; thickness of each layer tf = 0.166 mm. Normal beams were coded as BN-2, and beams with CFRP treatment were coded as BF-1. Installation of CFRP was carried out after the beam specimen reached 100% concrete strength. LVDT verti-cally on the right and left sides recorded the middle vertical deformation. Strain on tensile reinforcement and on compressed concrete fiber was recorded by strain gauge A, B, C, which was installed as shown in Figure 1.

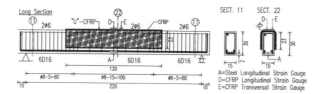

Figure 1.   Normal beam structure model (BN-2) without CFRP, beam (BF-1) treated with CFRP.

4   RESULTS AND DISCUSSION

Load testing on the BN-2 showed bending behavior similar to that of an over-reinforced beam. As shown in Figure 2, the destruction of concrete in the compressed area was accompanied by considerable deflection while the tensile reinforcement did not yield.

4.1   Load deflection

The destruction of concrete took place at peak loads of 292.25 kN with the deflection at 18 mm. The load test on the BF-1 test beam also showed bending behavior similar to that of an over-reinforced beam. Strengthening with CFRP showed better performance compared to normal beams without reinforcement. Figure 3 shows the peeling of the CFRP layer in the compressed area on the BF-1 treated beam due to continuous load increases.

This happened because the result of the pressure toward the transverse originating from the beam in the compressed area reached peak strain. Load was recorded at 386.88 kN and the deflection that occurred at the peak load was 26.72 mm. Figure 4 shows the relationship between load and deflection of both BF-2 and BN-1 beams.

The increase in load capacity can be seen from the curve shown in Figure 4. The magnitude of the normal BN-2 maximum beam deflection was 18 mm; the BN-2 beam

Figure 2.   Compression failure of normal beam BN-2.

Figure 3.   Compression failure of treated beam BF-1.

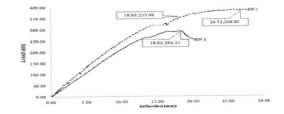

Figure 4.   Load–vertical deflection of BN-2 and BF-1.

can withstand loads up to 292.25 kN, while BF-1 can withstand loads of around 355 kN. Thus there is an increase in the strength of the load of (355−292.25)/292.25 x 100% = 21.47%.

4.2   Load–transverse concrete strain

In normal BN-2 beams, there is pressure on the concrete in the longitudinal direction, followed by a strain in the transverse direction. Normal BN-2 beams do not have stirrup reinforcement installed in the bending area, so the transverse strain is really only held by the strength of the concrete. The maximum strain of 0.001135 occurs at a load as large as that compressed by 292.25 kN. The wrapping treatment with CFRP on BF-1 beam serves to determine the extent of the role of restraint in increased beam flexural capacity. The CFRP wrapping serves as a restraint on the transverse strain (Hadi, 2009). It is expected to delay strain due to workload. The maximum strain of 0.006392 occurs at a load of 386.88 kN. This means that at this load, the beam is destroyed due to a compression failure, which is marked by rupture of the CFRP layer according to the transverse direction. From the results of testing for the two beams, it can be seen that the beam with CFRP treatment (BF-1) has the ability to withstand higher loads, and provides a greater strain value. The increase in expenses incurred is equal to (325.00−292.25)/292.25 x 100% = 11.21%.

4.3   Reinforced concrete beam ductility

Analysis of displacement ductility in normal beam elements showed vertical deformation (δy) of 13.50 mm with a first crack at 260 kN. When reaching its ultimate load, which is 292.25 kN, the vertical deformation (δu) reached 18.00 mm. Displacement ductility was calculated as: 18.00/13.50 = 1.33. The treated beam showed a vertical deformation (δy) of 14 mm at 315 kN. The ultimate load of 386.88 kN occurred at vertical deformation (δu) 26.72 mm. Therefore the displacement ductility was calculated as: 26.72/14.00 = 1.91. Results of analysis showed that the beam treated with CFRP had an increased ductility of: $\Delta \delta$ = (1.91−1.33)/1.33 x 100% = 43.14%.

4.4   Flexural moment capacity increase

Bending moment capacity on normal BN-2 beams was calculated based on the experimental test results that produced the maximum load for normal beams multiplied by the distance between the support and the load position: M_{BN2} = 0.5 P_{BN2} x Ls = 1.023 x 10^5 Nm.

For the treated beam, the ultimate bending moment that occurred was M_{BFI}= 1.354 x 10^5 Nm. Thus the moment increase is flexible: ΔM = (1.354−1.023)/1.023 x 100% = 32.38%.

4.5   Evaluation of crack patterns

Figures 2 and 3 depict the crack patterns of BN-2 and BF-1 beams, beginning at the middle span of the lower fiber by 53 kN of vertical hair cracks. The subsequent cracks occurred on the right and left sides of the first crack, with successive vertical directional at about 70 kN and 80 kN loads. With a load of 292.25 kN, there was a breakdown in the beam compression section. Results of the tensile reinforcement strain were 0.002022 smaller than the tensile yield strain of 0.002368. The crack pattern that occurred in the BF-1 beam started at a load of 68 kN with the appearance of hair cracks. Subsequent cracks also occurred on the right and left sides of the first crack with a directional orientation that tended to tilt at a load of 142 kN. At a load of 204 kN, the crack became longer. Furthermore, CFRP peeling first occurred at a load of 376 kN. Increasing the load to 386.88 kN damaged the beam compression section, and all CFRP bandages were exfoliated.

In reference to E. G. Nawy (1998) and ASCE Library and Ganga Rao (1998), these beams were categorized as "medium" slender beams, and the collapse generally was called "diagonal pull." The direction of the crack tilted closer to the angle 450 to the pedestal. This kind of crack pattern was illustrated as a variety of diagonal tensile collapse, which is assumed to have a shear effect on the beam.

5 CONCLUSION

The research concluded as follows: (a) The vertical deflection of the treated beam is small compared to the normal beam at the same load position. (b) Beams with CFRP treatment can withstand greater loads than normal beams without CFRP treatment. (c) Beams with CFRP treatment have increased ductility about 43.14%. (d) The transverse strain on treatment beam is small compared to normal beam strain value at the position of the load collapse. (e) The increase in flexural moment strength in beams with CFRP treatment is 32.38%.

REFERENCES

ACI Committee 440 2R-08. 2008. Guide for the design and construction of externally bonded FRP systems for strengthening concrete structures.
ASCE Library & Ganga Rao, P. V. V. 1998. Bending behavior of concrete beams wrapped with carbon fabric. *Journal of Structural Engineering, 124*(1).
E. G. Nawy, P. E. 1998. *Beton Bertulang-Suatu Pendekatan Dasar (terj. Bambang Suryoatmono)*. Bandung: Refika Aditama, Cet. Kedua.
Hadi, M. N. S. 2009. Behaviour of eccentric loading of FRP confined fibre steel reinforced concrete columns. *Construction and Building Materials, 23*(2), 1102–1108.
Hidayat, B. A. 2016. Studi Eksperimental Penggunaan Fiber Reinforced Polimer (FRP) pada Balok Kondisi Lentur dan Geser. Tesis, Magister Teknik Sipil, Universitas Diponegoro.
Pangestuti, E. K., & Handayani, F. S. 2009. Penggunaan Carbon Fiber Reinforced Plate sebagai Tulangan Eksternal pada Struktur Balok Beton. *Media Teknik Sipil, 9.*
Park, R., & Paulay, T. 1974. R*einforced Concrete Structure*. Christchurch, NZ: Wiley.
Sobuz, H. et al. 2012. Use of carbon fiber laminates for strengthening reinforced concrete beams. *International Journal of Civil and Structural Engineering, 2.*
Utami, S. R. L., Nuroji, N., & Antonius, A. 2016. Pengaruh Pembalutan Carbon Fiber Wrap (CFW) Terhadap Daktilitas Balok Beton Bertulang. *Informasi dan Ekspose hasil Riset Teknik SIpil dan Arsitektur, 12*(2), 140–155.

Engineering, Information and Agricultural Technology in the
Global Digital Revolution – Hendrawan & Wijayanti Dual Arifin (eds)
© 2020 Taylor & Francis Group, London, ISBN 978-0-367-33832-9

Flood mitigation of settlements on the sea tide with a water-filled rubber weir as an efficient innovation (a case study of flood in Semarang)

Fatchur Roehman
Sultan Fatah University, Demak, Indonesia

ABSTRACT: Water resources in Indonesia are diverse, but it is not easy to process them due to, for example, flooding and rob problems in Semarang. This research examined the hydraulic effectiveness of water-filled rubber weirs through a stability analysis of numerical models. The aim was to analyze the flow load stability as well as the efficient material characteristics. The research was carried out on weir prototypes in the laboratory with variations in the simulation of elevation models in the upstream water level (10 cm to 40 cm) and the simulation of downstream elevation (5 cm to 10 cm). The variables studied were weir weight, mud pressure, earthquakes, and hydro-statics. The laboratory tests revealed that the weight able to withstand the rolling force from upstream was 0.155 (Ton.Meter), with mud pressure of 0.003 (Ton.Meter), seismic force of 0.025 (Ton.Meter), and hydrostatic force of 0.084 (Ton.Meter). The conclusion was that the weir weight has a stabilizing factor. Hence, it can overcome flood mitigation by using efficient materials.

Keywords: Disaster Mitigation, Water-Filled Rubber Weir, Flood and Rob, Efficient

1 INTRODUCTION

The mitigation of floods and robs in Semarang needs to be handled with water-filled rubber weirs because the material used is an efficient technology (Ahmad, 2009). People can easily get such materials and participate in maintaining the weirs so that surrounding settlements and roads are not flooded (Roehman, 2017).

Flood control measures can include reducing flood peaks, flood walls, or closed channels, decreasing the peak surface of the flood by increasing the speed of water, repairing channels, transferring flood water through drainage, reducing runoff by processing land, and treating flood lands (Roehman, 2015). The purpose of this study was to analyze the effectiveness of hydraulic water-filled rubber weirs and their material characteristics, and to perform stability analysis on water-filled rubber weir models. The problem of this research was to study the hydraulic effectiveness of the water-filled rubber weir, to examine the characteristics of its material, and to conduct a stability analysis of the numerical model of the water-filled rubber weir (Roehman, Wahyudi, & Ni'am, 2019; Wahyudi, Ni'am, & Le Bras, 2012).

2 STUDY OF LITERATURE

Weir is a construction built across a river and intended to raise a river's water level upstream. Barrage can adjust the water level (Zhang, Tam, & Zheng, 2002). The rubber weir can be expanded and deflated as needed, and it can be filled with water or air (Whitworth et al., 2000).

Normative references in the planning of rubber weirs can use SNI 03-2415-1991 regarding the method of calculating flood discharge, SNI 03-1724-1989 concerning hydrological and hydraulic guidelines for buildings in rivers, and SNI 03-2401-1991 concerning weir planning procedures (Alhamati et al., 2005).

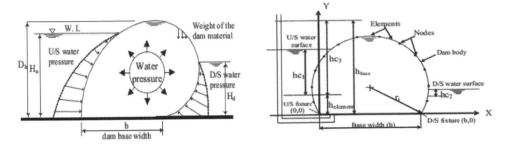

Figure 1.   The water pressure system in the weir (Alhamati et al., 2005).

Infrared weir is a flexible, foldable cylindrical inflatable structure made of rubber material attached to a rigid base inflated by air, water, or a combination of air/water (Alhamati et al., 2005; Plaut & Fagan, 1988).

Hydraulically, a weir must be able to serve the planned water level, open automatically if flooding exceeds a certain limit, and resist saltwater intrusion. Saltwater trapped upstream of the weir must be pushed downstream (Abdullah, 1999). A weir consists of a rubber sheet made from genuine or synthetic rubber that is elastic, strong, hard, and durable and reinforced with a nylon thread arrangement that provides tensile strength, as well as a synthetic rubber base material such as ethylene propylene diene monomer (EPDM) or chloroprene rubber (CR). The force can be calculated using the following approach (Alhamati et al., 2005).

The research was conducted in Semarang by simulating closed-circuit canals at the Civil Engineering Laboratory of Sultan Agung Islamic University, Semarang over four months.

3   RESEARCH CHARTS

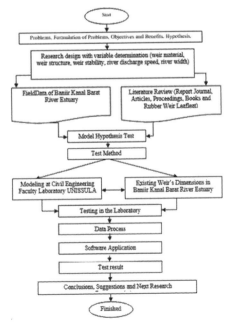

Figure 2.   Research flow chart (Source: Process in the laboratory, Fatchur Roehman Researcher, 2019).

4 RESULTS AND DISCUSSION

Figure 3. Triangle, quadrangle, cylinder, and trapezoid weirs (Source: Simulation in the laboratory, Fatchur Roehman Researcher, 2019).

Table 1. Weir's weight moment data on rolling force.

No	Elevation (m) Upstream	Downstream	Element	Υ water (Ton/m^3)	Width, B (m)	Height, H (m)
1	0.4	0.05	W1	1	0.6	0.4
2	0.4	0.1	W1	1	0.6	0.35
3	0.4	0.15	W1	1	0.6	0.3
4	0.4	0.2	W1	1	0.6	0.25
5	0.4	0.25	W1	1	0.6	0.2
6	0.4	0.3	W1	1	0.6	0.15
7	0.4	0.35	W1	1	0.6	0.1
8	0.4	0.4	W1	1	0.6	0.05

Form Factor	Volume m^3	Weir's Weight (Ton) (Ton)	Arm (m) x	y	Moment on Weir's Weight(Ton.M) Resistance	Rolling
0.5	0.072	0.072	0.3		0.022	
0.5	0.063	0.063	1.3		0.082	
0.5	0.054	0.054	2.3		0.124	
0.5	0.045	0.045	3.3		0.149	
0.5	0.036	0.036	4.3		0.155	
0.5	0.027	0.027	5.3		0.143	
0.5	0.018	0.018	6.3		0.113	
0.5	0.009	0.009	7.3		0.066	

(Source: Simulation results in the laboratory, Fatchur Roehman Researcher, 2019)

After eight simulations, as shown in Table 1, the biggest moment in the upstream elevation was 40 cm and the biggest moment in the downstream elevation was 0.25 cm. It produced a moment on the weir's weight of 0.155 (Ton.Meter).

The weir can withstand rolling forces. It can deflate both automatically and manually under the planned conditions. The weir body is safe against public disturbances and river transportation, and it is resistant to sediment abrasion and safe for water flow and sediment/garbage transport.

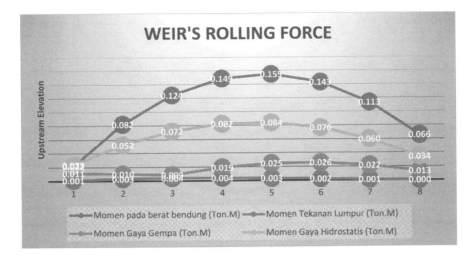

Figure 4. Weir's rolling force simulation results.
(Source: Processing of simulation results, Fatchur Roehman Researcher, 2019)

5 CONCLUSIONS AND SUGGESTIONS

The rubber weir functions to raise and lower the water level by expanding or deflating the weir body. The choice of material influences the flow's pressure stability. In Indonesia, water-filled rubber weirs have not been easy to obtain in the free market. The construction of water-filled rubber weirs utilizes geomembrane material as a form of efficient technology. Suggestions for further research include considering flow stability, the weight of the weir, and the carrying capacity of the soil. The hope is that researchers can further develop this flood control method so that weirs can be constructed without damaging the river ecosystem.

REFERENCES

Abdullah, D. T. 1999. *Bendung karet di Indonesia, makalah pada seminar bendung karet.* Serang.

Ahmad, I. A. 2009 Rubber dam usage for endodontic treatment: A review. *International Endodontic Journal, 42*(11), 963–972.

Alhamati, A. A. N. et al. 2005. Behavior of inflatable dams under hydrostatic conditions. *Suranaree Journal of Science and Technology, 12*(1), 1–18.

Plaut, R. H., & Fagan, T. D. 1988. Vibrations of an inextensible, air-inflated, cylindrical membrane. *Journal of Applied Mechanics.* American Society of Mechanical Engineers Digital Collection, *55*(3), 672–675.

Roehman, F. 2015. Delta preservation and handling diversion abrasion: Coastal model of the Sudeten Wulan River (case study in land arising in coastal Wedung, Demak, Central Java). *Proceedings of International Conference: Issues, Management and Engineering in the Sustainable Development of Delta Areas in Semarang, Indonesia.*

Roehman, F. 2017. The effectiveness of water-filled rubber weir management on flood mitigation caused by rain and "rob" (Java slang) in Demak District, Semarang, Indonesia. *Procedings of International Conference: Problem, Solution and Development of Coastal and Delta Areas, Semarang, Indonesia.*

Roehman, F., Wahyudi, I., & Ni'am, F. 2019. Analysis of physical model rubber weir containing water as motion weir for flood and rob handling. *International Journal of Civil Engineering and Technology (IJCIET), 10*(4), 219 227.

Wahyudi, S. I., Ni'am, M. F., & Le Bras, G. 2012. Problems, causes and handling analysis of tidal flood, erosion and sedimentation in northern coast of Central Java: Review and recommendation. *International Journal of Civil and Environmental Enginering, 12*(4), 65–69.

Whitworth, J. M. et al. 2000. Use of rubber dam and irrigant selection in UK general dental practice. *International Endodontic Journal, 33*(5), 435–441.

Zhang, X. Q., Tam, P. W. M., & Zheng, W. 2002. Construction, operation, and maintenance of rubber dams. *Canadian Journal of Civil Engineering.* NRC Research Press, *29*(3), 409–420.

Engineering, Information and Agricultural Technology in the Global Digital Revolution – Hendrawan & Wijayanti Dual Arifin (eds)
© 2020 Taylor & Francis Group, London, ISBN 978-0-367-33832-9

Weighting of the attribute importance weights using improved multiple linear regression and modified digital logic for process based fraud detection

Erba Lutfina & Safira Nuraisha
Universitas Semarang, Semarang, Indonesia

ABSTRACT: Fraud detection becomes one of the solutions to overcome the fraud case that occurred in banks. In the Fraud detection process, each PBF (Process Based Fraud) attribute has different effects to indicate fraud. The MDL (Modified Digital Logic) method is used to weight the PBF attributes. MDL method produces the attribute's importance weight that matches the impact of PBF attributes. However, the role of the expert is still very significant to assess the attribute's importance weight. This study aims to modify the weighting procedure of the attribute's importance weight in MDL method by adding Improved Multiple Linear Regression method (IMLR). By replacing the input previously given by the expert to the automatically weighting procedure. Then the results of both methods were evaluated using confusion matrix. Based on the experimental results, MLR method shows that the classification using all attributes importance weights has a better result with an accuracy of 99.5%.

Keywords: Fraud, Process Mining, Modified Digital Logic, Weight Of Attribute Importance, Regression

1 INTRODUCTION

Fraud has been the focus of research in recent years. (Aalst, Schonenberg & Song 2010) Fraud also caused banks to lose billions of dollars each year. Fraud detection methods becomes one of the solutions to overcome the fraud cases that occurred in the banking world. Fraud can occurs in the transaction data or in the running business processes. Data mining approach is used to analyze the transaction data, while the process mining approach is used to analyze the business processes. A process mining approach is used In this research, because the method proposed in this study is to analyze the business processes.

Fraud detection using data mining has been done in many studies. For example, using the Classification method, the Neural network algorithm, Self-Organizing Maps algorithm, empirical analysis, and web service collaboration method (Ramakalyani & Umadevi 2012). In addition, fraud detection using process mining is performed by (Sarno et al. 2015) using ARL hybrid algorithm and process mining and Fuzzy MADM. (Huda 2015) performed fraud detection using the process mining method. The study match the business rules with SOP to obtain PBF (Process Based Fraud) attributes. In the Fraud detection process, each PBF attribute has different effects to indicate fraud. To weight each PBF attribute, the MDL (Modified Digital Logic) method is used.

MDL method generates the attributes importance weights that correspond to the impact of the PBF attributes. However, in the MDL method the role of expert is still very significant to assess each attribute importance weights. This will be problematic if there is a change of research object or expert change. To overcome the problem of weighting attribute importance weights in MDL method, Improved Multiple Linear Regression (IMLR) method is proposed. IMLR is assisted by K-means and Moving Average (MA) techniques to overcome the influence of data outliers. This methods can be used to predict a value and find the coefficient weight of each attribute acording to (Sari, Ginardi & Suciati 2015). Prediction is obtained from multiplication of training data with regression coefficient. The regression coefficient shows the weight value of each attribute in the prediction process. The regression coefficient of each variables can be used as an input on the MDL method to overcome the weighting of attribute importance weights automatically.

Based on the problem analysis above, the weighting process of attribute importance weight can be solved by using regression coefficient. The regression coefficient is the result of the RLB process to calculate the attribute importance weight on the MDL method. The proposed method is expected to be used to calculate the attribute importance weight of PBF in determining the value of fraud detection.

2 METHODOLOGY

2.1 Dataset

In this research, data were collected from Process-Based Fraud detection system. The data are grouped into 806 cases for data testing and 300 cases for data training. The dataset consists of several cases that deviates from SOP. The deviations are identified based on the criteria of the PBF attribute. An example of a dataset is shown in the table below. In case 102, there is one violation of the skip sequence attribute and one in the wrong pattern attribute. Case 102 has a fraud detection value of 1, indicating that case 102 is fraud. Compared with case 200 which appears three times in the Throughtput time min attribute, the fraud detection value is only 0.878. The above example shows that each PBF attribute has different weights to generate fraud detection values.

2.2 Attribute PBF

Process-Based Fraud is a fraud that occurs because of a violation of business processes. In research conducted by (Huda & Ahmad 2014) mentioned some attributes of PBF to identify processes that violate the SOP. The PBF attribute is shown below:

Table 1. PBF attribute.

Attributes Name	Description
Throughput time min	The event execution time is faster than the standard event time
Throughput time max	The event execution time is slower than the standard event time
Skip sequence	An event is skipped from the actual flow
Skip decision	An event is skipped from the actual flow that have branches
Wrong duty decision	The originator executes more than one event in the event decision
Wrong resource	The event is executed by an unauthorized originator
Wrong pattern	The executed event does not match with the SOP sequence pattern
Wrong duty sequence	The originator executes more than one event in the event sequence
Wrong duty combine	The originator executes more than one event in the event sequence and decision
Wrong decision	Events that have executed decisions do not comply with the SOP
Event pararel	More than one event is executed simultaneously

2.3 Modified Digital Logic (MDL)

(Vats et al. 2014) Modified Digital Logic (MDL) is one of the most commonly used methods to estimate parameter weights. This method includes expert opinion to set priority of each parameter. Experts will define priority by assigning a value of 1 for less important parameters, value 2 for balanced priority and value 3 for more important parameters. Based on expert opinion, decision table is formed to show comparison of each parameter. After that, estimate the number of possible positive decisions with N = n (n-1)/n, where n is the number of attributes/parameters. P is the sum of all positive decisions for a given parameter and Wj is the final weight with equation:

$$Wj = \frac{Pj}{\sum_{j=1}^{n} Pj}$$

2.4 Improved Multiple Linier Regression (IMLR)

Improved Multiple Linear Regression method in this study is the MLR prediction method by building a gradual model in each linear regression function. This method is assisted by K-Means and Moving Average methods. The K-Means technique is implemented in selecting PBF attributes to form a pattern and calculation of the regression coefficient (Sari, Ginardi & Suciati 2015). The method is expressed as follows:

Table 2.   Improved multiple linear regression using K-means.

Algorithm 1. Improved Multiple Linear Regression using K-Means()
Start input history data as X, k, n init N as number of data object X, c, beta, y'; for i=i:N get X(1:n,:) calculate c = kmeans(X,k) Calculate beta using (2) Calculate y' using (1) endfor end

The impelentation of Multiple Linear Regression improved by Moving Average method is expressed as follows:

Table 3.   Improved multiple linear regression using moving average.

Algorithm 2. Improved Multiple Linear Regression using K-Means()
Start input history data as X, n init N as number of data object X, beta, y'; for i=i:N get X(1:n,:) Calculate beta using (2) Calculate y' using (1) Calculate y' using (5) endfor end

145

2.5 Classification

Determine the variables classification for each fraud detection value by using k-nn classification method. The classification result of fraud detection using the attribute importance weights compared with fraud detection classification result without using the attribute importance weights calculation. Comparison of classification results is done to compare the effect of importance weights attribute.

2.6 Validation

The result of this research is the comparison of fraud detection classification using the attribute importance weight. Testing of MLR method performed in the form of t test value, F test and R2 coefficient test. The evaluation process of classification results is done by calculating the accuracy value using confusion matrix method. TP (True Positive) and TN (True Negative) represent the correct classification while FP (False Positive) and FN (False Negative) represent the incorrect classification. The formula is shown as follows:

$$Accuracy = \frac{TP + TN}{TP + TN + FP + FN}$$

3 RESULTS AND DISCUSSION

The results conducted by using 806 data case from SOP deviation obtained the attribute importance weights. Attribute importance weights are calculated using the MDL method. Input on MDL method is weighting coefficient generated by IMLR method as follows:

Table 4. Weighting coefficient.

Attribute	Weight coefficient
Skip sequence (A1)	1
Skip decision (A2)	0,995
Throughput time min (A3)	0,375
Throughput time max (A4)	0,375
Wrong resource (A5)	0,976
Wrong duty sequence (A6)	0,962
Wrong duty decision (A7)	0,884
Wrong duty combine (A8)	0,919
Wrong pattern (A9)	0,467
Wrong decision (A10)	0,878
Event parallel (A11)	0,427

From the table above, the weight of each attribute then becomes input on the MDL method to generate the attribute importance weight. The comparison matrix of attribute weights is as follows:

The evaluation result using confussion matrix is as follows:

$$Accuracy = \frac{TP + TN}{TP + TN + FP + FN} = 241/242 = 0,995 = 99,5\%$$

The comparison result shows that k-nn classification with k = 3 value without using attribute importance weights has lower result with an accuracy value of 98.7% compared to the classification using all attribute importance weight with accuracy value of 99.5%.

Table 5. Comparison matrix of attribute weights.

	A1	A2	A3	A4	A5	A6	A7	A8	A9	A10	A11	weight
A1	2	3	3	3	3	3	3	3	3	3	3	0,132
A2	1	2	3	3	3	3	3	3	3	3	3	0,124
A3	1	1	2	3	1	1	1	1	1	1	1	0,058
A4	1	1	1	2	1	1	1	1	1	1	1	0,050
A5	1	1	3	3	2	3	3	3	3	3	3	0,116
A6	1	1	3	3	1	2	3	3	3	3	3	0,107
A7	1	1	3	3	1	1	2	1	3	3	3	0,091
A8	1	1	3	3	1	1	3	2	3	3	3	0,099
A9	1	1	3	3	1	1	1	1	2	1	3	0,074
A10	1	1	3	3	1	1	1	1	3	2	3	0,083
A11	1	1	3	3	1	1	1	1	1	1	2	0,066
Total Weight												1

Table 6. Confusion matrix result.

Classification Class	Prediction Class			
	Very Confident Fraud	Confident Fraud	Fraud	Not Fraud
Very Confident Fraud	37	0	0	0
Confident Fraud	0	0	0	0
Fraud	0	0	0	0
Not Fraud	1	0	0	204

4 CONCLUSION

From the results of research shows that the weighting of attribute importance weights by using Improved Multiple Linear Regression method on MDL method proved to be done automatically.

The results also show that Improved Multiple Linear Regression is suitable to produce the weight of each attribute. The results of the evaluation show that the classification using all attributes importance weights has a better result with an accuracy of 99.5%. This shows that all attributes importance weights affect the accuracy of the classification of fraud values. Future research is going to be developed further by using other independent variables or add independent variables that have a significant effect on the change of fraud value.

REFERENCES

Aalst, W.M.P. Van Der Aalst, Schonenberg, M.H. & Song, M. 2010, Time Prediction Based on Process Mining.

Huda, S. 2015, Fuzzy MADM Approach for Rating of Process- Based Fraud, no. November.

Huda, S. & Ahmad, T. 2014, Identification of Process-based Fraud Patterns in Credit Application, pp. 84–89.

Ramakalyani, K. & Umadevi, D. 2012, Fraud Detection of Credit Card Payment System by Genetic Algorithm, vol. 3, no. 7, pp. 1–6.

Sari, Y.A., Ginardi, R.V.H. & Suciati, N. 2015, Color Correction Using Improved Linear Regression Algorithm, no. March 2016.

Sarno, R., Dewandono, R.D., Ahmad, T. & Naufal, M.F. 2015, Hybrid Association Rule Learning and Process Mining for Fraud Detection, no. April.

Vats, S., Vats, G., Vaish, R. & Kumar, V. 2014, Selection of optimal electronic toll collection system for India : A subjective-fuzzy decision making approach, vol. 21, pp. 444–452.

Engineering, Information and Agricultural Technology in the
Global Digital Revolution – Hendrawan & Wijayanti Dual Arifin (eds)
© 2020 Taylor & Francis Group, London, ISBN 978-0-367-33832-9

Feasibility analysis of transferring Terboyo Terminal to Penggaron Terminal using a location theory approach

Widodo, M. Handajani & Ismiyati
Diponegoro University, Semarang, Central Java, Indonesia

ABSTRACT: The Semarang city government has designated Terboyo Terminal as a Type C Terminal, thereby making it into a parking facility for heavy vehicles. As a result of this policy, intercity bus services were moved from Terboyo Terminal to Penggaron Terminal. This study aimed to analyze the feasibility of transferring intercity bus services from Terboyo Terminal to Penggaron Terminal using a location theory approach. Descriptive analysis and analysis of user perceptions were carried out. The results of this study were: (1) Penggaron Terminal, in terms of the performance of its access road, has been crowded, and its location is far from other transit facilities, so its accessibility is poor. (2) From users' perception, reaching Penggaron Terminal is more difficult and less profitable compared to reaching Terboyo Terminal. It was concluded that the location of Penggaron Terminal is not suitable for use for bus transfers from Terboyo.

1 INTRODUCTION

Semarang is the largest trading city in central Java, located along the traffic corridors moving people and goods from Jakarta to Surabaya and vice versa, so it acts as the final node of the trip and as a transit node. Semarang accommodates various patterns of movement of people who have interests in the city and then return to their original location. Semarang also channels the movement of people whose final destination is outside Semarang after temporarily sheltering there. As described earlier, the role of the terminal is vital in the matter of accommodating these travelers.

The terminal serves not only as a transportation node but also as a place for loading and unloading passengers, the gathering of passengers and vehicles, resting, and even as a place to store or repair vehicles in the short term (Morlock, 1978). Since the terminal is one of the components of transportation and the place where various complex activities occur, it requires adequate space and a good location so that it does not cause traffic problems around it (Sihono, 2006).

Semarang previously had two Type A passenger terminals, i.e. Terboyo Terminal and Mangkang Terminal, and one Type B terminal in Penggaron. UU No. 23/2014 requires a division between regional and state authorities, including in terms of transportation. The authority of a Type A terminal falls to the state (Ministry of Transportation), Type B terminals come under the jurisdiction of the provincial government, and Type C terminals are governed by the regency/city government. Because in the case of transfer of terminals authority is related to regional assets, the state grants authority to the city government. The city government previously maintained all terminals, whether they were handed over to the state or the city government manage the terminals itself, as long as the terminals were classified as Type C. Therefore the Semarang city government handed authority over Mangkang Terminal to the state and ceded Penggaron Terminal to the provincial government, while Terboyo Terminal was not handed over to the state. The Semarang city government, through the mayor's Decree No. 551.22/1169 in 2016, determined Terboyo Terminal is a Type C terminal to be used as a heavy transport parking facility.

For the interests of the government, the transfer of terminals may be necessary. But what must be considered is how such a transfer affects passengers and vehicle operators. Criteria that need to be considered for passengers include consumer surplus, saving money and time, ease of moving to other modes of transportation, and ease of access. For operators, the criteria include operational costs to access the terminal location and the area for vehicle repairs (Apriyanto, 2009). Wicaksono explains that the transfer of bus terminals to suburban areas makes passengers uncomfortable because of the distance (cited in Eryana, 2002). This can cause the terminal to be quiet because passengers are reluctant to use it and prefer other places to take vehicles. Passengers' reluctance is also influenced by the purpose for their trip, the time it takes to reach their location, the proximity of the terminal to the place of origin, travel costs, and destination (Eryana, 2002).

This research was motivated by the fact that many bus passengers and operators were reluctant to move from Terboyo Terminal, so illegal activity has emerged there. This research should serve as an evaluation of the policies that have been carried out, and should help to formulate appropriate policies to overcome these problems.

2 METHOD

2.1 Data collecting method

The type of data in this research included secondary data and primary data. The secondary data were obtained from institutions, literature, newspapers, and other sources. The primary data were obtained from field surveys and questionnaires given to passengers and drivers around Terboyo Terminal. The respondents comprised 60 passengers and 40 drivers.

2.2 Data analysis method

The approach used in this study was a quantitative descriptive method used to identify the location of Penggaron Terminal from existing data and to analyze the results of questionnaires distributed to passengers and drivers. The descriptive method was used to examine the location of the Penggaron Terminal in terms of accessibility and road performance. User perception analysis was used to study what causes passengers and drivers to be reluctant to move to Penggaron Terminal. The questionnaire instrument was tested for validity and reliability, and the results were eligible for use.

3 RESULT

3.1 Accessibility of the terminal

When viewed from the position of the road, Penggaron Terminal is an off-street terminal because it is not located right on the edge of the highway. Penggaron Terminal is located approximately 600 meters from Brigjen Sudhiarto Road, which is a Class II Primary Collector's Road with Provincial Road status. This road is the main link in the flow of traffic from the city center to the outskirts of the southeastern region of Semarang (e.g., Gayamsari, Pedurungan), as well as the exit to several other areas such as Mranggen, Purwodadi, and Blora.

The road network in Semarang was developed with concentric radials, where Penggaron Terminal, which is located in the southeast, is not the main traffic meeting point. Penggaron Terminal is easier to reach by vehicles from locations in southeast Semarang, such as Purwodadi and Blora, but it is more difficult to access by vehicles or passengers coming from the east (e.g., from Demak, Kudus, Pati, and Rembang), and south (e.g., Magelang, Solo, and Salatiga), so it is quiet there.

Based on the analysis of the performance of the main road to Penggaron Terminal, i.e. Brigjen Sudhiarto, traffic from the direction of the city shows a high level of saturation (VCR > 0.75),

Table 1. Analysis of VCR and level of service of Brigjen Sudhiarto Road (from exit tolls to Penggaron).

Location	Time	V	C	VCR	LOS	Survey taken
Node 1	Morning	4,681.8	5,559.84	0.84	D	Wednesday
	Afternoon	4,498.2	5,559.84	0.81	D	
	Evening	4,924.2	5,559.84	0.89	E	
Node 2	Morning	4,705	5,957	0.79	D	Monday
	Afternoon	2,965	5,957	0.50	C	
	Evening	4,593	5,957	0.77	D	
	Night	2,104	5,957	0.35	B	
Node 3	Morning	1,342	3,675	0.37	B	Saturday
	Afternoon	3,006	3,675	0.82	D	
	Evening	2,838	3,675	0.77	D	
	Night	1,625	3,675	0.44	B	

Table 2. Linkages between Penggaron Terminal and other transit facilities.

Transit facility	Distance km	Reaching*	Cost Rp	Travel time hour
Harbor	18.5	2 times	3,500	2.33
		1 time	7,000	2.00
Airport	18.5	2 times	3,500	2.75
Railway Station	17.5	2 times	7,000	2.16
		1 time	3,500	1.75
Terminal A	26	1 time	3,500	2.58

* by using BRT

especially during peak hours in the morning and evening. Such conditions have the potential to cause congestion on the road, which of course will result in prolonging travel time for a bus to reach Penggaron Terminal. The road performance based on the results of a study conducted by Mudiyono and Anindyawati (2017) and on a study from Laporan Andalalin Transmart Penggaron (2018) are shown in Table 1.

The location of the terminal is ideally easy to use to access other transit facilities in the city. This will make it easier to move between modes so that users will be better able to save time and travel costs. The location of Penggaron Terminal is still considered unable to accommodate intermodal transfers with other transit facilities in the city of Semarang because it is far away, so it takes a long time to reach it. The linkage between the location of Penggaron Terminal and other transit facilities can be seen in Table 2.

3.2 User perception

From the survey results, it can be seen that several factors cause passengers to still use Terboyo Terminal.

3.2.1 Habit

Of the respondents, 42% were passengers who use the terminal more than twice a week, while 55% of the respondents use the terminal twice a week and the remaining 3% use the terminal fewer than twice a week. From this, it can be seen that many respondents are passengers whose daily activities are close to Terboyo Terminal so that the pattern of their trips has been formed there. Of course, it would be very difficult to move these passengers.

3.2.2 *Proximity of the location to passengers' origin*

The location of Penggaron Terminal is less strategic for passengers from the east. The survey results revealed that passengers with an eastward destination prefer using Terboyo Terminal over Penggaron Terminal. Distance also affected passengers' decisions when determining where to wait for transportation.

3.2.3 *Travel time*

The passengers were mostly students and workers who are required to be on time in starting their activities. This will influence their choice in where to wait for transportation.

3.2.4 *Travel cost*

The cost factor will affect passengers in determining where to wait for transportation. The majority of respondents who are students who use public transportation/buses will choose a place that can be accessed at a lower cost.

3.2.5 *Ease of reaching*

Terboyo Terminal's location is more strategic than Penggaron Terminal's location, making it easier to reach.

3.2.6 *Ease of getting transportation*

In terms of safety and comfort, at this time, the Terboyo Terminal location was very unsupportive as a place to wait for a bus. But the location is still busy as if guaranteeing that there will be convenient transportation.

The survey results from the drivers revealed some things that made them reluctant to move to Penggaron Terminal.

(1) The AKDP and AKAP bus services are separated.

Most drivers felt that the separation of AKAP bus services to Mangkang Terminal and AKDP bus services to Penggaron Terminal made it difficult for them to exchange passengers due to the distance between Penggaron Terminal and Mangkang Terminal.

(2) AKAP buses that are supposed to go to Mangkang Terminal are still operating around Terboyo Terminal.

According to the drivers, many buses still operate around Terboyo Terminal because they were left out by authorized officers. This is a cause of mistrust of the new rules.

(3) Road access is more difficult.

According to the drivers, the access road to Penggaron Terminal is more difficult because it is further from the toll gate, there are many red lights, and the traffic is crowded.

(4) Travel time is increased.

Most drivers consider that the travel time to Penggaron Terminal is longer due to more difficult access. Bus drivers from locations to the east such as Rembang, Pati, and Kudus, felt this especially hard because they have to go further to reach Penggaron Terminal.

(5) Rest time is reduced.

Most drivers consider that their rest time will be reduced because of the longer travel time to Penggaron Terminal.

(6) Income is reduced.

According to the drivers, the transfer to Penggaron Terminal has caused their income to decrease because passengers are reluctant and the cost of fuel is increasing.

4   CONCLUSIONS

The location of Penggaron Terminal is currently not suitable to be used as a place to transfer intercity buses from Terboyo Terminal. This is because (1) the accessibility of Penggaron Terminal is not good, as indicated by poor road performance and distance from other transit facilities. (2) Passengers are reluctant to move to Penggaron Terminal because the location of

Terboyo Terminal is more strategic and closer, so it is cheaper to use and easier to reach. Drivers are reluctant to move to Penggaron Terminal because the passengers are hesitant and access is difficult, resulting in increased travel time, increased fuel costs, and reduced driver rest time.

REFERENCES

Apriyanto, T. 2008. Kerangka Evaluasi Pengembangan Terminal Bus Antar Kota. *Jurnal Teknik Sipil dan Perencanaan*, 2(10) (July), 85–92.

Eryana, I. 2002. Aspek Kemudahan Pencapaian dalam Penentuan Lokasi Terminal Bus (Kasus Terminal Bus Terboyo Semarang). M.S. Thesis. Magister Teknik Sipil, Diponegoro University, Semarang.

Laporan Analisis Dampak Lalu Lintas Pembangunan Transmart Penggaron. 2018.

Morlock, E. K. 1978. *Pengantar Teknik dan Perencanaan Transportasi*. Jakarta: Penerbit Erlangga.

Mudiyono, R., & Anindyawati, N. 2017. Analisis Kinerja Ruas Jalan Majapahit Kota Semarang (Studi Kasus : Segmen Jalan Depan Kantor Pegadaian sampai Jembatan Tol Gayamsari). *Jurnal Smartcity*, 345–354. Retrieved from http://jurnal.unissula.ac.id/index.php/smartcity/article/download/1735/1302.

Sihono. 2006. Pengaruh Lokasi terhadap Aktifitas Terminal (Studi Kasus Terminal Giri Adipura dan Sub Terminal Krisak Kota Wonogiri), M.S. Thesis. Magister Teknik Pembangunan Wilayah dan Kota, Diponegoro University, Semarang.

SK Walikota Semarang Nomor 551.22/1169. 2016. Tentang Penetapan Terminal Terboyo sebagai Terminal Angkutan Jalan Tipe C dan Lokasi Parkir Angkutan Barang di Kota Semarang.

Undang-Undang No. 23 2014 tentang Pemerintahan Daerah.

Engineering, Information and Agricultural Technology in the
Global Digital Revolution – Hendrawan & Wijayanti Dual Arifin (eds)
© 2020 Taylor & Francis Group, London, ISBN 978-0-367-33832-9

Author Index